AF906076

SpringerBriefs in Political Science

SpringerBriefs present concise summaries of cutting-edge research and practical applications across a wide spectrum of fields. Featuring compact volumes of 50 to 125 pages, the series covers a range of content from professional to academic. Typical topics might include:

A timely report of state-of-the art analytical techniques
A bridge between new research results, as published in journal articles, and a contextual literature review
A snapshot of a hot or emerging topic
An in-depth case study or clinical example
A presentation of core concepts that students must understand in order to make independent contributions

SpringerBriefs in Political Science showcase emerging theory, empirical research, and practical application in political science, policy studies, political economy, public administration, political philosophy, international relations, and related fields, from a global author community.

SpringerBriefs are characterized by fast, global electronic dissemination, standard publishing contracts, standardized manuscript preparation and formatting guidelines, and expedited production schedules.

More information about this series at http://www.springer.com/series/8871

Luis Fernando Medina Sierra

Beyond the Turnout Paradox

The Political Economy of Electoral Participation

Springer

Luis Fernando Medina Sierra
Universidad Carlos III de Madrid
Getafe-Madrid
Spain

Universidad del Rosario
Bogotá
Colombia

ISSN 2191-5466 ISSN 2191-5474 (electronic)
SpringerBriefs in Political Science
ISBN 978-3-319-73947-2 ISBN 978-3-319-73948-9 (eBook)
https://doi.org/10.1007/978-3-319-73948-9

Library of Congress Control Number: 2017964722

© Springer International Publishing AG 2018
This work is subject to copyright. All rights are reserved by the Publisher, whether the whole or part of the material is concerned, specifically the rights of translation, reprinting, reuse of illustrations, recitation, broadcasting, reproduction on microfilms or in any other physical way, and transmission or information storage and retrieval, electronic adaptation, computer software, or by similar or dissimilar methodology now known or hereafter developed.
The use of general descriptive names, registered names, trademarks, service marks, etc. in this publication does not imply, even in the absence of a specific statement, that such names are exempt from the relevant protective laws and regulations and therefore free for general use.
The publisher, the authors and the editors are safe to assume that the advice and information in this book are believed to be true and accurate at the date of publication. Neither the publisher nor the authors or the editors give a warranty, express or implied, with respect to the material contained herein or for any errors or omissions that may have been made. The publisher remains neutral with regard to jurisdictional claims in published maps and institutional affiliations.

Printed on acid-free paper

This Springer imprint is published by the registered company Springer International Publishing AG part of Springer Nature.
The registered company address is: Gewerbestrasse 11, 6330 Cham, Switzerland

Preface

A few years ago, I found myself in a social gathering with some old friends in Bogota. The conversation drifted past the usual political topics, about which we were in broad agreement, and led us to realize that we all had something else in common: we were all first-generation Bogotans whose parents had come to the city as youngsters when their families, from different regions of the country, fled the political violence of the 1940s and 1950s. Some, the majority at the meeting, were Liberals harassed by Conservative militias, but in some cases, the roles of party labels were switched. The question of what would have happened, had all of them lived in the same town, is one that I found a bit too eerie to ask. We were all, so to speak, grandchildren of *La Violencia*, the seemingly generic name that for Colombians denotes that viciously murderous period of "partisan cleansing." But at the same time, we were all children of the National Front, products of a system of higher education, both public and private, much larger than anything the country had known, fed by parents who put food on the table sometimes through public-sector employment, sometimes through protected domestic manufactures, always as part of the urban middle class that thrived under the import substitution of the 1960s and 1970s. Having grown in such an environment, we had done something that was unthinkable to our grandparents: we had more than once chosen to vote differently from what our parents did.

From the social scientist's point of view, this insignificant vignette concerning some middle-class, middle-aged Colombian urbanites is worth mentioning precisely because of how unremarkable it is. Colombia is not the only country whose citizens' political choices are, somehow, a response to broad changes in the environment where they live. Electoral maps often tell the story of their respective countries. The story of European industrialization can be elicited from the swaths of "red districts" that often corresponded to the grimy industrial belts of many of their capitals. The agrarian crises that blighted the American prairie in the late nineteenth century left their marks in the form of Populist bastions that largely disappeared from sight during the early 1900s. In country after country after country, the ballot boxes, filled with the outcome of millions of individual, largely decentralized choices, tell a story of shared life experiences shaped by geography, economic transformations, technological progress, and social mores, among other forces.

This simple and admittedly trite observation forms the starting point of my research agenda on voting. I believe that no theory of voting can claim to be satisfactory unless it illuminates how and why the citizens' voting choices respond to the social environment where they live. This may seem a truism but, as I will argue later, it already decouples my own program from other developments in the literature.

This book is the result of combining this starting point with a more controversial methodological choice. In the following pages, I will rely extensively on rational-choice theory. In short, my purpose in this book is to develop a theory of voting that puts the tools of game theory into the service of a structural view of electoral phenomena. In that sense, this book is something of a companion to my earlier work, *A Unified Theory of Collective Action and Social Change*, where I took a similar stance with respect to collective action problems which are, after all, closely related to voting.

Defending thoroughly such methodological choice would require an entire volume, and perhaps even that would not be good enough for some critics. So, I shall just limit myself to observing that if we want to understand the interaction between economic processes and voting, we need a theoretical framework in which individuals are able to understand at least a modicum of their economic environment and respond accordingly to it. There is room for disagreement about what we should expect from such a theoretical framework, but, at a minimum, rational-choice theory seems as good a candidate as any other paradigm to fulfill its requirements.

Rational-choice theory is at the heart of contemporary economic science. There is much to be improved in economics, but it is hard to deny that most of the advances in our current understanding of, say, inflation, trade, and business cycles have occurred within the idiom of rational choice. At the very least, this suggests that there are pragmatic reasons to adopt rational-choice theory as the language to integrate economics and politics within a single conceptual matrix.

These remarks, however brief and cavalier, ought to suffice here. In the end, the research program I am defending here stands or falls on the merits of its results. The reader is entitled to know from the start that in this book I will fall far short from a comprehensive theory of voting that meets the standards I just set out. At most, I will offer what I believe are useful first steps. But, as I will illustrate later, these first steps already allow us to make progress in the study of some specific aspects of voting, heretofore unexplored.

To some extent, this is an old research program, and one could say that it has largely been completed. There already is an abundant literature on voting from a rational-choice perspective that traces the impact of economic phenomena over voting outcomes. The works of Roemer (2001a) and Acemoglu and Robinson (2006) are but a few recent examples. This literature, however, has been unable to deal with a central aspect of any democratic political process: electoral turnout.

It has been widely accepted both by friends and critics of rational-choice theory that there is a problem in trying to explain why rational individuals would vote in the first place. If individuals are rational, so the argument goes, in the sense of basing their decisions on an assessment of the costs and benefits attached to them, they will find that there is no reason to vote at all because the impact of their individual vote, except in an artificially small electorate, is negligible.

Curiously, it is not easy to evaluate how damaging a problem this is. At first glance, it may seem, and it seems to many critics, that a theory that cannot explain why people vote is built on sand and is, therefore, inconsistent, unable to yield any useful products, and undeserving to be taken seriously. Rationalists take a different view. While acknowledging that this is a source of embarrassment, this has not stopped them from producing an entire theoretical edifice of political economy. Arguably, even if one does not have a satisfactory explanation of why people vote, it still is possible to have a good theory of how they vote when they do.

Subject to some caveats, I find this latter argument somehow convincing. But there is one reason why I believe this way of quarantining; as it were, the paradox of voting is damaging to the enterprise of rational-choice theory.

Turnout is one of the defining features of any democracy. It is not clear whether a specific level of turnout is an indicator of good or bad health for a democratic system, but there is no question that if we want to understand the choices open to organized collectivities in a competitive political system, we need to understand what makes some people vote and others not. Even a cursory reading of, say, American political events, to focus on a particularly low-turnout democracy, shows that the US political parties constantly try to understand what drives citizens to the polls; their electoral fortunes, and with them the political course of the country, depend on that.

By the standards I laid out at the beginning, a good theory of electoral turnout ought to elucidate the structural forces behind it. Just as we can trace the way in which the social environment affects how the citizens vote, we should also be able to trace how that same environment leads some other citizens to sit out the election. Electoral abstention, I firmly believe, is every bit as subject to influence from the socioeconomic context as voting. During the 1946 election that pushed Colombia one decisive step toward the abyss of *La Violencia*, electoral turnout reached unprecedented levels in a logic reminiscent of the spike in turnout that marked the fateful German election of 1933. Electoral participation in the USA follows a well-known and allegedly disquieting pattern: citizens with lower income, and lower educational levels that belong to specific ethnic minorities vote at lower rates than the average.

This is the point where I part company with the existing rationalistic theories of voting. For all their accomplishments, and there are many, I do not find them adequate for the purpose of analyzing the structural determinants of electoral participation. A common trait of all these approaches in their attempts at explaining voting has been to invoke the existence of components that modify the cost-benefit calculation making it rational for some citizens to vote. Civic duty (Riker and Ordeshook 1968), expressive value (Schuessler 2000), and ethical dispositions (Feddersen and Sandroni 2006) are some of the elements that have been brought out in such quest. There is no question that these are important considerations and that they all have the potential of leading to a richer theory of voting. But in their current state, they are not useful for my analysis because we do not know at this point how these subjective traits respond to the objective facts of the socioeconomic structure. Does income inequality promote civic duty? Are workers in the export sector more prone to expressive behavior than those in non-tradeables? Are citizens more likely to follow ethical dispositions in times of high inflation? The fact that these questions sound somewhat off-key indicates that a structural account of electoral turnout must start somewhere else.

To me, the right starting point is a proposition that many readers might find startling and even preposterous: from a game-theoretic point of view, voting poses no paradox. This flies in the face of a large body of scholarship, and so I will devote Chap. 1 to substantiate it.[1] In that chapter, I do not "solve" the paradox of voting but, rather, dissolve it. That is, I argue that it is largely an amalgam of ambiguities and unnecessarily restrictive assumptions.

A game-theoretic analysis of voting, without any ad hoc modifications of the basic cost-benefit analysis, can offer a satisfactory account of electoral participation. Such is the main lesson of Chap. 1. In itself, I believe that this is an important result, but it is not enough for my purposes. My goal is to understand how structural forces shape the patterns of turnout in an electorate. Thus, in Chaps. 2 and 3, I show how, once we overcome the paradox of voting, game theory can help us to establish very precise and intuitive patterns of electoral participation. Ultimately, the conclusion of those chapters might seem somewhat disappointing because they establish what the calculus of voting meant to prove from the very beginning: that the decision to vote responds to costs and benefits, just like any other decision.

But this initial disappointment should be ephemeral because, once we realize that voting is subjected to clearly discernible forces, a new viewpoint emerges. Instead of asking ourselves why people vote, it is possible to address questions about the comparative statics of electoral turnout. Chapter 4 exemplifies the kind of research that becomes feasible with these new results. In it I show how the operation of the welfare state determines the benefit structure of voting and, through it, the populace's electoral participation. As will become clear in that chapter, this approach can generate new testable hypotheses about turnout and its connections to a polity's socioeconomic environment. This chapter is, however, only an example of the type of research program I want to advance. Its full development is beyond the scope of this book, but given the advances so far, it is possible to discern the contribution such a research agenda would make to our understanding of democracy. Chapter 5 takes up these issues and offers a summary and concluding remarks.

Getafe-Madrid, Spain Luis Fernando Medina Sierra

[1]The analysis I present here builds on results I have obtained and reported elsewhere, in particular, Medina (2011).

Contents

Chapter 1
Beyond the Voters' Paradox

The life of the rational-choice theory of electoral participation resembles at times that of Viscount Medardo de Terralba, the fictional character of Italo Calvino's *The Cloven Viscount*. In Calvino's novel, the viscount bravely goes to battle in the frontlines only to be horribly wounded by a cannonball that splits him vertically in half. From that point on, each half, with one arm, one leg, one eye, one nostril and half a mouth, lives on its own. They never meet again, except for the final and felicitous showdown, but before that, their paths cross so that invariably one half sabotages the plans of the other.

In the case of rational-choice theory, the so-called "turnout paradox" was the cannonball that split what promised to be a unified research agenda into two separate programs, often at odds with each other: the theory of turnout and the spatial theory of electoral competition. The original program, as articulated for instance in Anthony Downs's classic work *An Economic Theory of Democracy* (Downs 1957) intended to develop the microfoundations of the calculus of voting and non-voting for individual citizens, then aggregate the resulting decisions and show how they would interact with the strategic decisions of political parties. The result would have been similar to the general equilibrium models in economics where the consumers' decisions are the basic input of the demand function and the firms' decisions are aggregated into the supply function.

This program was frustrated by the turnout paradox, that is, the result according to which if citizens respond rationally to their individual preferences and the costs and benefits of voting, then they should not vote. If the turnout paradox is correct, then there are no individually rational voting decisions to aggregate in the first place.

So, instead of one unifying treatment of electoral behavior and parties' choices we have two theories that have to rely on strange crutches, that rarely meet and that whenever they do, create trouble for each other. First we have a theory of electoral turnout that tries to explain why people vote if voting is costly but that has little to say about how those voting decisions depend at all on the citizens' preferences

© Springer International Publishing AG 2018

L. F. Medina Sierra, *Beyond the Turnout Paradox*, SpringerBriefs in Political Science,
https://doi.org/10.1007/978-3-319-73948-9_1

between candidates. Then we have a spatial theory of how parties compete for votes but that assumes away the turnout problem postulating that voting is costless.

The theoretical challenge is then, just like with the Viscount, to stitch the two halves together so that they no longer need to rely on these ad hoc limitations. In this book I will show how this is possible and offer a sketch of how this reunited theoretical body might move once its limbs are working together. In the resulting framework, each half will forsake some of the prosthetics that helped it come this far. The micro-level calculus of voting will no longer need to ignore the role of the political parties' strategies.

A full discussion of these points will have to wait until later chapters in the book. In the meantime, the first order of business is to evaluate the patient (or should it be half-patients?) to determine the type of intervention needed and its chances of success.

The first thing to notice, the central point of this chapter is that, unlike Viscount Medardo's case, the turnout paradox was a self-inflicted wound resulting from an incorrect assessment of the properties of the rational-choice theory of voting. The rather peculiar history behind this dates to the early 80s and deserves to be told in detail not only for how odd it is, but also because it helps us see under a new light the recent progresses obtained in this matter. Let me first sketch the plot and the details will be filled in the rest of the chapter.

The turnout paradox was first stated by Downs in his foundational work when he carried out an analysis of the calculus of voting within the framework of decision theory. That is, he considered only the cost-benefit analysis of a single individual without factoring in the interactions of that individual's decisions with those of others. In doing so he was adhering to the highest standards of analysis of his time: game theory was then in its embryonic stage. By the early 80s, after the turnout paradox had already acquired the status of received wisdom, a small set of scholars realized that the decision-theoretic framework was flawed and set out to revisit the issue but this time using the tools of game theory. Ledyard (1984), Grofman and Owen (1984) and, more exhaustively, Palfrey and Rosenthal (1983) examined the properties of a game theoretic model of voting. The third of these contributions finally established an unequivocal conclusion: models of voting that considered the strategic interactions among agents admitted of a plethora of solutions many of them with very high degrees of electoral participation. In other words, there was no turnout paradox.

That result should have put the issue to rest. Remarkably, two years after their breakthrough, Palfrey and Rosenthal (1985) retracted in an exemplar display of intellectual honesty. They argued that the high-turnout equilibria they had obtained were in fact "fragile" (in ways that will be discussed below) and that hence, in their own words, "we have come full circle and are once again beset by the paradox of not voting" (Palfrey and Rosenthal 1985, p. 64). Such critique carried enormous weight. Not only it was developed by the authors of the previous result but also they were, already by then, some of the foremost scholars in this area. Thus died a short-lived enthusiasm and the turnout paradox regained its canonical status, if anything firmer than before.

But the retraction was wrong. As I will show in this chapter, in their self-criticism Palfrey and Rosenthal made some subtle mistakes. If those mistakes are corrected it becomes clear that the arguments about fragility are, depending on the actual meaning of the word "fragile," either non-issues or concerns of much lesser importance than what seemed at first glance.

To be sure, those game-theoretic models of the first generation have problems that need to be addressed. While successful at predicting high turnout, some of their other empirical predictions are unrealistic, their mathematical structure is unwieldy and some of the results lack a straightforward interpretation. Fortunately, there are already some recent developments in the literature that show how to build better models. Those will be the subject of the next chapter.

1.1 The Analysis of Voting in Decision Theory

What we now know as the rational-choice approach to voting began with Downs (1957) attempt at representing an individual voter's decision with simple mathematical expressions that, so the hope was, would enable a more precise analysis of the functioning of a democratic system. This is not the place to discuss the plausibility of that goal or the philosophical and methodological foundations of the enterprise. But, at least by the standards of the debates that would erupt afterwards, Downs's starting point was relatively non-controversial.

His basic contention was that individuals would vote based on a rational analysis of their interests and he was hardly alone in this assumption. At the time of Downs's writing, the notion that underlying interests, especially economic, would drive the citizens' voting behavior was already accepted by many as standard. Within the Marxist tradition, the debates around the electoral viability of social-democratic parties were, in a way, debates around the potency of the proletariat's economic interests in shaping its political choices. During the inception of the social democratic movement, those leaders who argued that electoral victory was imminent given the quantitative growth of the proletariat (as Przeworski and Sprague (1986) documents), were implicitly assuming that the workers' electoral choices would reflect their interests as economic agents. In a different though related intellectual tradition, Lipset built his oeuvre by tracing how economic interests would map into the fortunes of the political parties in democracies (e.g. Lipset 1960).

Not every political scientist subscribes to the tenets of methodological individualism and there is ample room to disagree about the way in which citizens form their views of what are their "true" interests. But the assumption that, whatever those interests are, individuals use them as a basis to make their electoral choices is hardly a shocking claim.

Understanding that electoral choices are plagued by uncertainty, Downs relied on expected-utility theory to build his formalism. This choice proved to be of great significance because it has been faulted, incorrectly as I will later show, with being responsible for precipitating the paradox of voting. It should be noted, however, that expected-utility theory is not an eccentric choice. Warts and all, it is still the dominant paradigm in economic analysis and, at the time Downs developed his theory, it was virtually uncontested.

Within the framework adopted by Downs, when facing choices under uncertainty, rational agents evaluate the alternatives according to the criterion of expected utility. Suppose that the act of voting has a cost of $c > 0$ and that there are two political alternatives on offer, x_L and x_R. A citizen i must decide whether or not to vote for her preferred candidate, say x_L. Her preferences for these candidates are represented by the utility function u_i so that, for instance, if candidate x_L emerges victorious, her utility is $u_i(x_L)$. According to the standard analysis of decisions under uncertainty, she will vote if the expected value of doing so exceeds the expected value of abstaining. In principle, the only benefit of voting, compared to abstaining, is the increase in the probability of victory of the favorite candidate. Let V_i and A_i denote the choices of voting and abstaining, $P(x|S_i)$ denote candidate x's probability of victory if agent i chooses strategy S_i and let w represent the agents' expected-utility function. Then, we can represent this analysis formally with the following inequality:

$$
\begin{aligned}
w(V_i) &\geq w(A_i) \\
P(x_L|V_i)u_i(x_L) + P(x_R|V_i)u_i(x_R) - c &\geq \\
P(x_L|A_i)u_i(x_L) + P(x_R|A_i)u_i(x_R) & \\
(P(x_L|V_i) - P(x_L|A_i))(u_i(x_L) - u(x_R)) &\geq c \qquad (1.1)
\end{aligned}
$$

Inequality 1.1 is the cornerstone of the decision-theoretic analysis of voting. In a simplified election between two candidates, decided by majority rule, the term $P(x_L|V_i) - P(x_L|A_i)$ is the probability that citizen i's vote will be pivotal. According to a well-known argument, this probability is vanishingly small. So, if we approximate it to 0, the result is that $w(V_i) < w(A_i)$. In other words, a rational citizen should not vote. This is, in essence, the paradox of voting.

Before studying the substantive issues, let's begin by noticing that the term "paradox" here does not denote a logical conundrum, like, say, the Liar's Paradox. The conclusion that voting contradicts rationality rests on a specific assessment about the relative magnitudes of its parameters. The term $P(x_L|V_i) - P(x_L|A_i)$ is not exactly zero, regardless of the number of voters. Conventionally we approximate it to zero but the reasons for doing so are not related to the formal structure of the underlying theory. In fact, the approximation $w(V_i) \approx 0$ is not entirely legitimate because we first need to determine the value of the other term, $u(x_L) - u(x_R)$. If we observe citizen i voting, the preceding analysis could lead to three possible conclusions: either she is violating the principles of rationality or her cost of voting is very low or she perceives a difference between the candidates so large that she considers it justified to vote regardless of the low chance of being pivotal. Conceivably, she regards candidate x_R as the devil incarnate or she is overestimating the value of $P(x_L|V_i) - P(x_L|A_i)$. Just because an agent is mistaken about the magnitude of a variable, does not mean that she is irrational. The preceding analysis does not necessarily imply that voting is paradoxical.

The paradox of voting is, then, an issue of measurement, not logic. Whenever we say that voting is not rational, we are really saying that we do not believe that the orders of magnitude of the terms $u(x_L) - u(x_R)$ and $P(x_L|V_i) - P(x_L|A_i)$ stack up in a way that justifies voting. But to defend such a view we should offer estimates of their relative magnitudes something that is rarely done. There has been some work on determining the size of pivot probabilities which is, as we shall soon see, a task fraught with difficulties. But this is only part of the argument. A full-blown statement of the paradox of voting should also pronounce on the magnitude of $u(x_L) - u(x_R)$.

Aside from its implications about an agent's decision to vote, the calculus of voting also determines the comparative statics of turnout. In fact, this is the pillar of our subsequent analysis. If we look past the paradox, Inequality 1.1 generates hypotheses about what determines an agent's voting decisions.

Suppose that for some citizen the expected utility of voting is lower than that of abstaining. Increases in the value of the inequality's left-hand side or decreases in its right-hand side will bring her closer to satisfying the condition that will lead her to vote. If we aggregate across agents, this means that increases in $u(x_L) - u(x_R)$ and decreases in c will increase the level of turnout.

Remarkably, although in this book I will study several variants of this model, for the most part significantly more complex, these comparative statics results will not change. In a way, it is possible to interpret this book's first part as an attempt to protect these results from the paradox of voting.

1.1.1 Departures from the Orthodox Model

Once the turnout paradox came to be accepted as a problem for rational-choice models of voting, it gave rise to significant efforts that, in spite of their differences, had in common their willingness to modify at least some of the key principles of the underlying model of human decision-making. There are many good reasons to do this. Even the most obdurate believer in rational-choice theory is willing to concede that it is only an approximation of the way individuals choose in real life. Some of the developments that have come out of this approach have become significant contributions to the modern theory of voting and as such need to be engaged here even though the approach I will take is different.

One of the first proposed solutions to the paradox was to postulate a term of "civic duty" that, allegedly, compensates the voter for the cost c (Riker and Ordeshook 1968; Tullock 1971). If we suppose that a voter experiences some additional utility d from the act of voting, resulting from the psychological benefit of fulfilling one's own obligations, then Inequality 1.1 becomes $[P(x_L|V_i) - P(x_L|A_i)][u_i(x_L) - u(x_R)] +$

$d \geq c$. Evidently, even allowing for $P(x_L|V_i) - P(x_L|A_i)$ to be arbitrarily close to 0, if $d > c$ citizen i will vote.

It is reasonable to assume that citizens who vote, often do so driven by some psychological motivation akin to that of civic duty. Thus, a general model of voting should, at least in principle, be able to accommodate this phenomenon.

A vast literature has endeavored to develop this idea further by, for instance, modifying the payoff functions by introducing expressive benefits (Schuessler 2000) or altering the incentive structure (Morton 1991, e.g.). Kanazawa (1998) obtained a model with high turnout by changing the way in which voters compute probabilities. Ample empirical and experimental evidence indicate that, indeed, many citizens are driven to vote by a sense of civic duty. That is one of the key lessons from studies on the effects of social pressure and appeals to altruism such as those of, for instance, Gerber et al. (2008) and Fowler (2006). But it is difficult to transform the original suggestion of Riker and Ordeshook into a fully operational research program inasmuch as civic duty remains there as a mere exogenous variable. In the limit, civic duty could end up offering a rather tautological explanation of voting without any relevant comparative statics results that connect turnout to structural features of the polity. After all, the term d is not connected to the candidates' properties and, instead, represents a strictly private psychological disposition beyond the scope of our analysis. If civic duty is the one and only factor that explains the citizens' decision to vote, all the model can tell us is that some citizens will vote for some unobservable reasons and some will not.

Recently, Feddersen and Sandroni (2006) and Feddersen et al. (2009) have proposed an approach that endogenizes the role of civic duty by endowing the agents, or at least some of them, with rule-utilitarian preferences and the accompanying reasoning abilities. In these models, some agents in the game derive some additional utility from following an ethical rule that maximizes social welfare, a move that is akin to the rather heterodox notion of "team reasoning" (Bacharach 2006) in which players base their actions on the best interests of the team to which they belong. Certain experiments have been able to confirm that, indeed, players use such criteria in coordination games (Mehta et al. 1994). Alternatively, this approach can be seen as related to the work on coalitional strategies where players' deviations can be collective and not merely individual as in the standard Nash equilibrium approach. An idea with a long pedigree in game theory, dating back to Aumann (1959), Bernheim et al. (1987), recent results suggest it could also be applied to large voting games (Ambrus 2006).

Another suggestive departure from the canonical rational-choice model is the one proposed by Jonathan Bendor, Daniel Diermeier, David Siegel and Michael Ting in their joint work (Bendor et al. 2003, 2011) which the authors adequately call a "behavioral theory." This approach, instead of postulating some ethical disposition that remains unexplained, acknowledges that individuals follow decision rules that are not necessarily optimizing and endeavors to extract from this assumption conclusions relevant for aggregate voting phenomena. Such work follows clearly on the footsteps of Herbert Simon's notions of "bounded rationality," in particular the idea that, instead of looking constantly for the best possible choice out of a possibly highly complex choice set, individuals save in their costs of information-gathering and -processing by sticking to a "satisficing" choice only to revise it if significant discrepancies arise between outcomes and aspirations (Simon 1957).

The resulting framework allows the authors to study jointly voting behavior and partisan strategies, much in the unifying spirit I defend in this book. As is often the case in social sciences, there is, however, a trade-off between descriptive realism and simplicity. Work on satisficing rules requires intensive use of numerical simulations, something that makes it at times difficult to interpret the results and assess their generality. As long as the turnout paradox retains the central place it has been afforded in the literature, any future research would have no choice but to accept these costs under any circumstance. If, as I will show later, the turnout paradox is not the forbidding result it is considered to be, then it is possible to obtain some analytic results on political competition that analyze simultaneously the choices of voters and parties. In Chap. 4 I will illustrate this with one example. Of course, such simplicity also comes at the expense of descriptive realism; no real-life voter is a perfect optimizing machine. But I believe it is salutary for the advance in this area to have different tools, some more accurate but complex, some more simple but tractable, so that the right balance can be found in each specific application.

1.2 The Strategic Analysis of Voting

The original calculus of voting, as formulated by Downs, is couched in terms of decision theory. It assumes that, in making their voting decisions, citizens do not take into account each other's choices. At first glance this may seem a natural assumption but it leaves ill-specified several aspects of the model. A crucial quantity in the equation, $P(x_L|V_i) - P(x_L|A_i)$, cannot be determined by relying only on decision theory because the probability that a vote is pivotal depends on the likelihood with which the remaining citizens vote. For instance, if nobody else is voting, one vote is decisive. A citizen's decision whether or not to vote depends on her assessment, however capricious, of what the other citizens are doing.

Scholars working within the framework of decision theory soon recognized this and proposed a solution: to interpret the term $P(x_L|V_i) - P(x_L|A_i)$ as an expectation. So, if the agents expect the election to be close, they will be more likely to vote because they will perceive their own probability of being pivotal as high enough to compensate for the cost c.

Although not entirely rigorous, this formulation is not at all bad. It served to generate a large literature on the connection between turnout and perceptions of closeness.[1] The general formalism that I will develop later on will arrive at a similar

[1] An early exponent of this literature, closely linked to the calculus of voting studied here, is the contribution by Ferejohn and Fiorina (1974). Cox (1988) offers an illuminating discussion of the methodological issues involved.

position. But there are some loose ends that need tightening. In particular, it is vulnerable to a problem familiar in game theory: the agents' expectations could potentially be self-defeating in a way that invalidates the original prediction. To illustrate, suppose that most citizens expect the election to be very close and, therefore, are prone to vote. But if this expectation is widely shared, this may lead to the election no longer being close. As the citizens realize this, some of them may come to the conclusion that it will not be worth to vote. But then, these decisions may increase the probability of a vote being pivotal so that, it may be worth voting after all. Rational agents would then be caught in a loop of vacillations where once they make a decision, they realize that it is no longer optimal.

There is no guarantee that things will happen this way as expectations could be self-reinforcing. At any rate, we need to develop a formal treatment of voting that allows us to discern how serious this problem is in each particular case. We need to analyze voting from the point of view of game theory.

While in decision theory we analyze an individual agent's problem in isolation from that of others, in game theory we regard all the decisions in a group as interdependent. As a result, instead of looking for decisions that are individually optimal, we need to look for groups of decisions that are all in equilibrium, that is, that are optimal against each other.

We need some extra concepts and notation to capture the additional complexity of this kind of analysis. From now on, we shall define a voting game as a combination of players, strategies and payoffs. Formally, a voting game $\Gamma = \langle N, \{S_k\}_{k=1}^N, \{\hat{u}_k\}_{k=1}^N \rangle$ is a triple such that:

- N is the set of players (citizens).
- There are two candidates in the election, x_L, x_R which may be either concrete individuals or policy packages. $\mathcal{O}$ defines a set of possible outcomes of the election. In its most general form, this set can be different from the set of candidates and later I will consider different possible sets.
- Each citizen k has a strategy space represented by S_k. Each individual strategy space is a set defined as $S_k = \{A_k, V_k^L, V_k^R\}$ where A_k stands for "Player k abstains" and V_k^M stands for "Player k votes for candidate M." We will soon see that, given their preferences, citizens can be assigned to groups of supporters of one candidate or another. When that happens, one of their voting strategies will be weakly dominated because it will mean voting for a candidate the voter deems inferior and therefore can be eliminated from the strategy space. As a result, often we can avoid a proliferation of superscripts, whenever it does not lead to confusion, by identifying the players' strategies simply as S_k omitting the candidate they support. For each player, the full strategy space will be $\Delta(S_k)$, that is, the set of all the linear combinations between A_k and V_k. Intuitively, this means that we will allow for the possibility of randomized choices where a player uses each strategy only with some probability less than 1. The parameter $\sigma_k \in [0, 1]$ will denote any such mixed strategy. So, if we say that player k chooses strategy σ_k we are saying that she votes with probability σ_k and abstains with probability $1 - \sigma_k$. The set $S = \times_{k=1}^N S_i$ represents the set of all the possible strategy profiles.

- The citizens' preferences are represented by Bernoulli utility functions that assign real numbers to elements in the outcome set, depending on the strategy chosen by the player. Formally, $\hat{u}_k : \mathcal{O} \times_{k=1}^{N} \Delta(S_k) \rightarrow \mathfrak{R}$.
- The candidates partition the set N in two subsets. Subset N_L, the set of supporters of x_L is defined as $\{i : \hat{u}_i(x_L, s_i) > \hat{u}_i(x_R, s_i)\}$ for any s_i and, conversely, subset N_R, the set of supporters of x_R is $\{j : \hat{u}_j(x_R, s_j) > \hat{u}_j(x_L, s_j)\}$ for any s_j. For the sake of completeness, we could define also a set N_0 of indifferent voters N_0 as $\{k : \hat{u}_k(x_L, s_k) = \hat{u}_k(x_R, s_k)\}$ for any s_k but, for the most part, there is no loss of generality in assuming this to be an empty set. Given this partition, I will use subscript i to denote a supporter of x_L, j to denote a supporter of x_R and will reserve the subscript k to the cases where it is necessary to refer to both types of players at once.
- If player k belongs to set N_M, $S_k = \{A_k, V_k^M\}$ is the strategy space of player k.
- An outcome function $g : \Delta(S) \rightarrow \mathcal{O}$ determines, for each strategy profile among all the citizens, the outcome of the election.

This setup still does not include uncertainty, something essential for our analysis. We need to be able to calculate expected utilities so, let's define a probability distribution over all the possible strategy profiles $\mu : S \rightarrow [0, 1]$. Later this distribution will be a very convenient way of generalizing our equilibrium analysis. In the meantime, suffice it to say that, since μ represents a probability, it is subject to the typical constraints of never assigning a nonsensical, negative probability to any profile and of assigning probabilities such that the sum total across profiles is equal to 1. Mathematically, $\mu(s) \geq 0, \forall s \in S$ and $\sum_{s \in S} \mu(s) = 1$. Sometimes we will be interested in events that are more complex than a single strategy profile and so we will need some additional notation for them. I shall denote the probability of event E under the measure μ as $P(E|\mu)$. When considering strategy profiles restricted to a subset M of agents, I will use the notation μ_M. The most common cases where this will happen will be when considering just one agent's actions, say μ_k, or the actions of all the players except k (μ_{-k}). With these elements we can define each player's expected utility given by the von Neumann-Morgenstern utility functions $w_k : \times_{k=1}^{N} \Delta(S_k) \rightarrow \mathfrak{R}$ such that:

$$w_k(\mu) = \sum_{s \in S} \hat{u}_k(g(s), s_k)\mu(s).$$

Often, but not always, we will be interested in a special type of probability distribution. Suppose an arbitrary profile of actions $s = \{s_1, s_2, \ldots, s_N\} \in S$. We shall say that the probability distribution $\sigma : S \rightarrow [0, 1]$ is a Nash-play distribution if $\sigma(s) = \sigma(s_1)\sigma(s_2)\cdots\sigma(s_N) \equiv \sigma_1\sigma_2\cdots\sigma_N$. In other words, under the distribution σ, each player's strategy is independent from the strategy of the remaining players. Technically speaking, while the general distributions μ capture all the correlated equilibria of the voting game, which will be of interest later on, the Nash-play distributions σ will help us characterize the game's Nash equilibria which play a prominent role in the literature on voting games.

The following assumption about u_k does not imply any loss of generality and, instead, greatly simplifies calculations. From now on, I will assume that $\hat{u}_k(g(s), s_k)$ is separable, that is, can be expressed as $\hat{u}_k(g(s), s_k) = u_k(g(s)) + c_k(g(s), s_k)$. The function u_k will represent the way in which the policy outcome affects the citizen's welfare by virtue of her ideological preferences. In contrast, the function c_k represents the cost that the act of voting entails for the citizen. Later we will see that there might be good reasons to make this cost term depend on the policy outcome so that, to keep the analysis as general as possible, here I allow c to depend on $g(s)$ but, following the standard procedure, I will analyze instances where it does not.

All the models that I will discuss in this book are particular cases of this one. In this chapter we will study different variants of this basic model. A detailed analysis of these models will allow us to go beyond blanket statements about the paradox of voting by ascertaining the degree to which it afflicts any specific model. As we will see, once we conduct this analysis, the scope of the paradox turns out to be much smaller than what the literature assumes. Moreover, those models pervaded by the paradox are excessively restrictive and, ultimately, uninteresting. Once we relax some of their assumptions, the paradox of voting becomes a rather peripheral concern.

1.3 The Pure Majority-Rule Game

The simplest voting game, the one that has largely remained as the default case of analysis, is the pure majority-rule game with homogenous players. In this game the outcome is defined by simple majority and there are only two types of payoff functions. For the sake of analytical precision, we will denote this game as Γ_{PM}. Since in this game the vote tally will play a prominent role, we need a special item of notation for it. From now on, the vote tally of candidate M is denoted by $T_M = \#\{k : s_k = V_i^M\}$. To refer to candidate M's vote tally excluding the action of his supporter k, I will use T_M^k. So, the formal definition of Γ_{PM} is a voting game that, in addition to the elements introduced above, also fulfills the following conditions.

- The outcome set is $\mathcal{O} = \{x_L, x_R\}$.
- The outcome function is defined by:

$$g(s) = \begin{cases} x_L \text{ if } T_L > T_R, \\ x_R \text{ if } T_R > T_L, \\ \mathcal{L} \text{ if } T_L = T_R, \end{cases} \tag{1.2}$$

  where $\mathcal{L}$ is a lottery that gives equal odds to each candidate.
- The utility functions are such that for every i in N_L and every j in N_R:
  - $\hat{u}_i(x_L, A_i) = u_L(x_L)$; $\hat{u}_j(x_R, A_j) = u_R(x_R)$,
  - $\hat{u}_i(x_L, V_i) = u_L(x_L) - c$; $\hat{u}_j(x_R, V_j) = u_R(x_R) - c$,

- $\hat{u}_i(x_R, A_i) = u_L(x_R)$; $\hat{u}_j(x_L, A_j) = u_R(x_L)$,
- $\hat{u}_i(x_R, V_i) = u_L(x_R) - c$; $\hat{u}_j(x_L, V_j) = u_R(x_R) - c$.

This is the benchmark case in the literature. Historically it was the first model studied as a strategic voting game and remains the source of most of the analysis even to this day. Beginning with the pioneering study by Ledyard (1984), game theorists quickly realized that the results pertaining to the paradox of voting had a very different character in a game-theoretic framework than the one they had within the earlier decision-theoretic paradigm. To appreciate this, notice that, in general, universal abstention cannot be a Nash equilibrium. If no one else votes, an individual citizen will find it worthwhile to vote: at the small cost of c, she can decide the entire country's government. By itself, this should count as circumstantial evidence that the reach of the paradox is narrower than what we usually think.

Building on this approach, Palfrey and Rosenthal (1983) and, independently and almost simultaneously, Grofman and Owen (1984), obtained the first breakthrough toward understanding the true scope of the paradox. They proved that in the voting game just described, with two candidates and majority rule, there are several equilibria, many of them with high levels of turnout.

This is a very important result that deserves to be studied closely. Just like the authors do, for the time being, I will analyze this model under the assumption of Nash play so that I will restrict my attention to strategy profiles σ where the probability of any individual action is independent of the rest. To calculate the Nash equilibria of this game we need to obtain the players' best-response correspondence. For an arbitrary citizen in N_L, taking as given the remaining players' strategy profile σ_{-i}, the expected payoffs of the strategies are:

$$\begin{aligned} w(A_i, \sigma_{-i}) &= (u_L(x_L) - u_L(x_R))P(T_L^i \geq T_R + 1|\sigma_{-i}) \\ &\quad + \tfrac{1}{2}(u_L(x_L) - u_L(x_R))P(T_L^i = T_R|\sigma_{-i}) + u_L(x_R), \\ w(V_i, \sigma_{-i}) &= (u_L(x_L) - u_L(x_R))P(T_L^i \geq T_R|\sigma_{-i}) \\ &\quad + \tfrac{1}{2}(u_L(x_L) - u_L(x_R))P(T_L^i = T_R - 1|\sigma_{-i}) + u_L(x_R) - c. \end{aligned}$$

Let the term $u_L(x_L) - u_L(x_R) \equiv \delta_L(x_L, x_R)$. So, from these payoffs we obtain player i's optimal strategy $\sigma_i^*(\sigma_{-i})$, expressed by the following best-response correspondence:

$$\sigma_i^*(\sigma_{-i}) = \begin{cases} 1 & \text{if } P(T_L^i = T_R|\sigma_{-i}) + P(T_L^i = T_R - 1|\sigma_{-i}) > \frac{2c}{\delta_L(x_L, x_R)}, \\ [0,1] & \text{if } P(T_L^i = T_R|\sigma_{-i}) + P(T_L^i = T_R - 1|\sigma_{-i}) = \frac{2c}{\delta_L(x_L, x_R)}, \\ 0 & \text{if } P(T_L^i = T_R|\sigma_{-i}) + P(T_L^i = T_R - 1|\sigma_{-i}) < \frac{2c}{\delta_L(x_L, x_R)}, \end{cases} \quad (1.3)$$

with an entirely analogous best-response correspondence for members of N_R. In what follows I will refer to the quotient $2c/\delta_L(x_L, x_R)$ (or $2c/\delta$ for short) as the cost-benefit ratio for voter i.

Intuitively, this best-response correspondence ratifies what we already know from the decision-theoretic analysis: a citizen will vote only if the expected benefit, in this case, the value of deciding the election times the probability of doing so, exceeds the cost of voting. Seemingly, this would mean that this analysis is doomed to reproduce the paradox of voting. For any large electorate, the probability of casting the decisive vote, it could be argued, is exceedingly small.

But this last step is not a correct reasoning; the pivot probability is not even a meaningful concept unless we describe the underlying strategy profile. As a matter of fact, from a game-theoretic standpoint, the pivot probability is endogenously determined as a function of the game's payoffs as we shall soon see.

This game, Palfrey and Rosenthal noticed, may have a plethora of equilibria, many of them with high levels of turnout. This may seem mysterious at first glance but stops being so once we keep in mind the underlying nature of an equilibrium. In decision theory the costs and benefits provide enough grounds to justify an agent's choice. But in game theory, if we want to make sense of a player's choice we need to place such choice in the context of a strategy profile including all the other players. To fix ideas, let's see some numerical examples.

1.3.1 Examples with Small Electorates

Since our ultimate purpose is to develop a theory adequate to the study of large elections, examples with a small number of voters may seem useless. But they will help us fix ideas and expose patterns that will reappear later in the more complicated cases. Readers who are not seasoned game theorists but do want to follow the details of the game-theoretic analysis later will benefit from studying closely these examples.

For simplicity's sake, apart from all the previous assumptions, I will also assume throughout that the value of the cost-benefit ratio is $2c$ (that is, $\delta = 1$ for every player). Later, once we are interested in the details of the voters' preferences, I will relax this assumption.

A 2-Voter Electorate

This is the simplest possible example. As it happens, its properties do not generalize well. It produces one type of equilibrium that will be rather exceptional in the more general framework. But it is helpful to study it however briefly precisely because it helps us place that exceptional equilibrium in context.

Let's assume that $N_L = \{1\}$, $N_R = \{2\}$ so that Player 1 belongs to N_L and Player 2 to N_R. Intuitively, it is clear that both players face a pivot probability of 1; if the other player votes, their vote can force a tie and if the other player does not vote, their vote can win the election. Formally:

$$
\begin{aligned}
P(T_L^1 = T_R|\sigma_{-1}) + P(T_L^1 = T_R - 1|\sigma_{-1}) &= 1 - \sigma_2 + \sigma_2, \\
&= 1. \\
P(T_R^2 = T_L|\sigma_{-2}) + P(T_R^2 = T_L - 1|\sigma_{-2}) &= 1 - \sigma_1 + \sigma_1, \\
&= 1.
\end{aligned}
$$

The equilibrium of this game depends on the value of c. If $c > 1/2$, then $2c > 1$ so that the pivot probability cannot be bigger than twice the cost-benefit ratio. In that case, the optimal strategy for both players is to abstain ($\sigma_1 = \sigma_2 = 0$). Instead, if $c < 1/2$, then their only optimal strategy is to vote ($\sigma_1 = \sigma_2 = 1$).

Each of these carries a lesson. The first case shows that for any voting game, a cost-benefit ratio above 1/2 guarantees that universal abstention is the only possible equilibrium. This is because the pivot probability, by definition, is bounded between 0 and 1. If $2c > 1$ then it is not possible to satisfy the condition for voting in the player's best-response correspondence. This rather trivial point has often led to damaging confusions as will become apparent later.

The second case is a rarity not only because of its turnout level of 100% but, what is conceptually more important, because in it both players choose a pure-strategy in equilibrium. As we will soon see, games within this framework only have pure-strategy equilibria whenever the two blocks of voters are of identical size (i.e. $N_L = N_R$). Additionally, in every equilibrium some players will be using mixed strategies. As is often the case, the use of mixed strategies in equilibrium can be difficult to interpret so this should serve as a call for extra caution. That said, this issue becomes irrelevant after the first stages of the analysis; it disappears in a more general framework.

A 3-Voter Electorate

The next simplest case possible is, of course, the one with three players. It turns out that adding one extra player significantly changes the pattern so that now the game has multiple equilibria, something that will become a recurring theme throughout this book.

Let $N_L = \{1, 2\}$ and $N_R = \{3\}$. From the standpoint of a player in N_L, say Player 1, the pivot probability is as follows:

$$
\begin{aligned}
P(T_L^1 = T_R|\sigma_{-1}) + P(T_L^1 = T_R - 1|\sigma_{-1}) = \sigma_2\sigma_3 + (1 - \sigma_2)(1 - \sigma_3) + \\
(1 - \sigma_2)\sigma_3.
\end{aligned}
$$

In other words, Player 1 is pivotal only in three possible events: both Players 2 and 3 vote, neither Player 2 nor Player 3 votes and Player 2 abstains while Player 3 votes. An analogue expression represents Player 2's pivot probability.

As for Player 3, her pivot probability is:

$$P(T_R^3 = T_L|\sigma_{-3}) + P(T_R^3 = T_L - 1|\sigma_{-3}) = (1-\sigma_1)(1-\sigma_2) + (1-\sigma_1)\sigma_2 + (1-\sigma_2)\sigma_1.$$

That is, Player 3 is pivotal only in any of the following events: neither Player 1 nor Player 2 votes, Player 1 abstains and Player 2 votes or Player 2 abstains and Player 1 votes.

Let's now look at the different possible equilibria of the game:

Pure-Strategy Equilibria We already know that if $c > 1/2$ then the game has only one equilibrium: nobody votes. But the other case is more interesting. As already stated, if $c < 1/2$ this game does not have any equilibrium in which all the players choose a pure strategy. It would be tedious to analyze all the possible combinations but, for the purposes of illustration, we can see how a putative equilibrium where only player 3 votes would break down.

To that end, conjecture an equilibrium profile $(\sigma_1^* = 0, \sigma_2^* = 0, \sigma_3^* = 1)$. If we substitute for these values in the pivot probabilities we obtain:

$$\begin{aligned}
P(T_L^1 = T_R|\sigma_{-1}) + P(T_L^1 = T_R - 1|\sigma_{-1}) &= 1, \\
&> 2c, \\
P(T_L^2 = T_R|\sigma_{-2}) + P(T_L^2 = T_R - 1|\sigma_{-2}) &= 1, \\
&> 2c, \\
P(T_R^3 = T_L|\sigma_{-3}) + P(T_R^3 = T_L - 1|\sigma_{-3}) &= 1, \\
&> 2c.
\end{aligned}$$

These values are incompatible with the postulated equilibrium for players 1 and 2. The conjectured profile had them choosing $\sigma_i = 0$ but, given that same conjecture, their pivot probability is larger than $2c$ which means that their optimal choice would be $\sigma_i = 1$. This contradiction makes the equilibrium unravel. A similar procedure could be used to verify all the other combinations.

Mixed-Strategy Equilibrium The game has an equilibrium where every player randomizes and these randomizations are mutually consistent because they make every player indifferent between her pure strategies. So, to calculate this equilibrium all we have to do is to solve the system of simultaneous equations that equates all the pivot probabilities to the cost-benefit ratio $2c$. The reader can verify that the following values are such solution:

$$\sigma_1^* = \sqrt{1-2c},$$
$$\sigma_2^* = \sqrt{1-2c},$$
$$\sigma_3^* = 1-\sqrt{1-2c}.$$

Pure-Mixed Equilibria In addition to the previous equilibrium, the game has two other equilibria. In these, one player uses a pure strategy and the other two choose a mixed one. They are:

$$\sigma_1^* = 1,$$
$$\sigma_2^* = 1-2c,$$
$$\sigma_3^* = 2c,$$

and

$$\sigma_1^* = 1-2c,$$
$$\sigma_2^* = 1,$$
$$\sigma_3^* = 2c.$$

For illustrative purposes, let's verify that the first one is, indeed, an equilibrium. This profile generates the following pivot probabilities:

$$P(T_L^1 = T_R|\sigma_{-1}) + P(T_L^1 = T_R - 1|\sigma_{-1}) = 4c(1-2c),$$
$$> 2c,$$
$$P(T_L^2 = T_R|\sigma_{-2}) + P(T_L^2 = T_R - 1|\sigma_{-2}) = 2c,$$
$$P(T_R^3 = T_L|\sigma_{-3}) + P(T_R^3 = T_L - 1|\sigma_{-3}) = 2c.$$

These equations validate the players' choices prescribed by the strategy profile postulated. That is, the strategy profile is an equilibrium. The same procedure can be applied to the second strategy profile with identical results.

There are several important conclusions to be extracted from this example. But before discussing them, let's study the more complicated case of a 5-player game.

A 5-Voter Electorate

The complexity of voting games grows very quickly with their size. In fact, since every best-response correspondence can be satisfied in three different ways, the number of possible equilibria that needs to be verified is $3^{N_L+N_R}$. So, whereas in the previous game there were 27 possible cases (out of which only 3 turned out to be equilibria), in this case an exhaustive computation would need to consider

243 cases. Checking all these cases would be exceedingly tedious and ultimately unnecessary for our purposes. Our goal is not to solve exhaustively a symmetric 5-player voting game with no practical relevance whatsoever but rather to find here insights and patterns that will be useful later on.

Let $N_L = \{1, 2, 3\}$ and $N_R = \{4, 5\}$.

Let's start with the expressions for the pivot probabilities. Since the game is symmetric, we only need to characterize the pivot probability of Player 1 (a representative member of N_L) and of Player 4 (a representative member of N_R). For Player 1, such probability would be:

$$\begin{aligned} &P(T_L^1 = T_R|\sigma_{-1}) + P(T_L^1 = T_R - 1|\sigma_{-1}) = \\ &\quad \sigma_2\sigma_3\sigma_4\sigma_5 + (\sigma_2(1-\sigma_3) + (1-\sigma_2)\sigma_3)(\sigma_4\sigma_5 + \sigma_4(1-\sigma_5) + (1-\sigma_4)\sigma_5) + \\ &\quad (1-\sigma_2)(1-\sigma_3)(\sigma_4(1-\sigma_5) + (1-\sigma_4)\sigma_5 + (1-\sigma_4)(1-\sigma_5)), \end{aligned}$$

Whereas for Player 4, it would be:

$$\begin{aligned} &P(T_R^4 = T_L|\sigma_{-4}) + P(T_R^4 = T_L - 1|\sigma_{-4}) = \\ &\quad \sigma_1(1-\sigma_2)(1-\sigma_3) + (1-\sigma_1)\sigma_2(1-\sigma_3) + (1-\sigma_1)(1-\sigma_2)\sigma_3 + \\ &\quad (1-\sigma_1)(1-\sigma_2)(1-\sigma_3)(1-\sigma_5) + \\ &\quad (\sigma_1\sigma_2(1-\sigma_3) + \sigma_1(1-\sigma_2)\sigma_3 + (1-\sigma_1)\sigma_2\sigma_3)\sigma_5. \end{aligned}$$

With these expressions at hand, we can proceed to calculate the game's equilibria. I will not show all of them, but will focus on some important and illustrative cases.

Pure-Strategy Equilibria Again, if $c > 1/2$ the game has a unique equilibrium where nobody votes. But if $c < 1/2$, just as in the previous case, there is no equilibrium in which all the players choose a pure strategy.

Mixed-Strategy Equilibrium The equilibrium in mixed strategies can be calculated by solving the system of five equations in five unknowns that results from choosing the case of indifference in all the best-response correspondences. Since the game is symmetric, we only need to find two values $p = \sigma_1^* = \sigma_2^* = \sigma_3^*$ and $q = \sigma_4^* = \sigma_5^*$ such that:

$$\begin{aligned} P(T_L^1 = T_R|(p,q)) + P(T_L^1 = T_R - 1|(p,q)) &= 2c, \\ p^2q^2 + 2p(1-p)(2q(1-q) + q^2) + (1-p)^2((1-q)^2 + 2q(1-q)) &= 2c, \\ P(T_R^4 = T_L|(p,q)) + P(T_R^4 = T_L - 1|(p,q)) &= 2c, \\ (3p(1-p)^2 + 3p^2(1-p))q + (1-p)^2(1-q)^2 + 3p(1-p)^2(1-q) &= 2c. \end{aligned}$$

Unfortunately, obtaining an analytical solution of this system would be exceedingly unwieldy. But it can be solved for any numerical value of c and, as we will see later, we can obtain comparative statics results even without knowing the exact solution.

Pure-Mixed Equilibria This game has several such equilibria but, for illustrative purposes, I will describe two of them. Apart from that, one of these two will feature prominently later.

In the first equilibrium, the strategy profile is: $(\sigma_1, \sigma_2, \sigma_3, \sigma_4, \sigma_5) = (1, 1, 1 - \sqrt{c}, \sqrt{c}, \sqrt{c})$. In other words, two voters from the majority party vote with absolute certainty while the remaining three players randomize their choice. Notice that this is an asymmetric equilibrium because here Player 3 has the same payoff function as Players 1 and 2 and yet chooses a different strategy.

The second equilibrium is one where: $(\sigma_1, \sigma_2, \sigma_3, \sigma_4, \sigma_5) = (1 - \sqrt{1-2c}, 1 - \sqrt{1-2c}, 1 - \sqrt{1-2c}, 1, 1)$. In this other equilibrium it is the members of the minority, this time all of them, the ones that vote with certainty while the members of the majority randomize. Palfrey and Rosenthal identified this equilibrium, which I will denote from now on as the **P-R equilibrium**, as the one with the highest possible turnout. A 5-player voting game is the smallest game where such an equilibrium appears but it exists for large electorates and it plays a very important role in the study of voting games.

What Have We Learned from These Examples?

It may seem that there is little of interest to be learned, from the standpoint of a general theory of electoral turnout, by studying games with two, three or five players. But some of the results obtained here are more general and by studying them carefully we can clarify some of the confusions that over time led to incorrect notions about the "turnout paradox." Here I will simply enunciate the main lessons but a full statement of their implications will have to wait.

Multiple Equilibria The first thing to be noticed is that voting games have multiple equilibria. Not only that, the amount of equilibria grows at an alarming rate as the electorate becomes larger. Such multiplicity will complicate the subsequent analysis. Soon we will see that some minor and plausible changes in the underlying model lead to patterns that are easier to describe. But the multiplicity of equilibria will remain and we will need to develop different analytical tools to handle it.

High Turnout Small voting games have equilibria with high turnout. That is hardly surprising. The received wisdom about the "turnout paradox" claims that this is to be expected because the effect of the vanishingly small pivot probabilities only kicks in with larger electorates. But I will presently show that this is largely mistaken. The P-R equilibrium, which generates very high participation rates, not only exists in small games like the 5-player game just studied, but also in arbitrarily large games, albeit under some special conditions to be clarified.

Costs and Benefits Matter In all the equilibria obtained, the strategy profile is a function of the cost-benefit parameter. The preceding examples illustrate this point in two ways. First, the value of c determines whether it is possible to have an equilibrium with positive turnout. We already saw that when $c > 1/2$ no player votes. In small games this is a trivial point. Instead, in large games the equivalent result is more complicated. An accurate calculation of the values of c compatible with high turnout is essential to understand the true extent of the turnout paradox. Second, the examples above show that the equilibrium strategies of the voters are themselves a function of the cost-benefit structure of the voting game. This insight will be the cornerstone for the analysis of the political economy of electoral participation because it will help us understand how changes in the environment of an election, that is, changes in the parameters faced by the citizens, their costs and benefits of voting, affect their aggregate patterns of turnout.

1.3.2 *High Turnout in Large Games*

Let's now turn to the case of large games. To that end, we can begin with a numerical example that will help us fix ideas. For this example, let $N_L = 10^7$ and $N_R = 9 \times 10^6$, an electorate of 19 million citizens, large by most standards. Let the cost-benefit parameter be such that $2c = 4.2 \times 10^{-3}$. In principle, we should already be able to observe the effects of vanishing turnout at that size. Then, if we can obtain equilibria with very high electoral participation, that should militate strongly against the turnout paradox.

The first thing to notice is that this game has billions of equilibria (and even that may be undercounting). With 19 million best-response correspondences, the amount of strategy combinations that can satisfy them is beyond description.

Although there is no way to show all the game's equilibria, it might be useful to construct a few of them. Let T_L and T_R be the total number of players, call them "active players," for both parties L and R that will participate with some positive probability, possibly even with probability one. We stipulate that the remaining players will simply abstain. Let's assume, further, that $T_L = T_R + 1$.

From the standpoint of any active player $i \in T_L$, the following condition pertaining her pivot probability ensures that she will choose a mixed strategy q:

$$q^{T_R} + T_R q^{T_R - 1}(1 - q) = 2c.$$

By the same token, the pivot probability for an active player in T_R such that she votes is:

$$\frac{(T_R + 1)T_R}{2} q^{T_R - 1}(1 - q)^2 + (T_R + 1)q^{T_R}(1 - q) > 2c.$$

Although, absent numerical values for the parameters, we cannot know the exact solution to the first equation, it is easy to show that it is $q \approx T_R/(T_R + 1)$ which, for large T_R approaches 1. So, for such values of q, the two conditions are consistent.

From the point of view of the inactive players in N_L, the equilibrium condition on the pivot probability is:

$$\frac{(T_R+1)T_R}{2}q^{T_R-1}(1-q)^2+(T_R+1)q^{T_R}(1-q)>2c.$$

The most notorious one, especially for having the highest possible turnout is the P-R equilibrium where every citizen in N_R votes and every citizen in N_L randomizes. In this particular case, with the given parameters, a randomization $\sigma_i = 0.89989$ satisfies the conditions. This equilibrium profile produces an aggregate turnout of roughly 18 million voters, or, in percentage terms, 94.7%.

But there are many others. For example, we can have equilibria where the pattern is shifted so that now all the citizens in N_R randomize and some citizens in N_L vote while the remaining abstain. Consider the case where four million citizens in N_L vote and the nine million citizens in N_R randomize with $\sigma_j = 0.44433$. Now turnout is approximately equal to 13 million, a percentage of 68.4%.

Notice that this set of conditions describes not just one equilibrium but myriads of them. In fact, since we are analyzing a symmetric game where all players' payoffs are identical, and since these conditions do not specify which players will do what, then any selection of four million citizens out of N_L can constitute an equilibrium. So, to be more accurate, we have just added $\binom{10^7}{46}$ equilibria.

There are many other similar equilibria. For instance, an almost identical randomization will support equilibria where all the citizens in N_R randomize, 3,999,999 citizens in N_L vote and the remaining 6,000,001 abstain. There is another set of equilibria where all citizens in N_R randomize with $\sigma_j = 0.44422$ and 3,999,000 citizens in N_L vote while the remaining 6,001,000 abstain.

Following this same reasoning, we could add more and more astronomic amounts of equilibria with high turnout with the most diverse rates of participation. This is not a matter of being fastidious; as we shall see later, a very important confusion in the literature turns on exactly this point.

These high levels of turnout are no accident. It is easy to prove that a P-R equilibrium exists for any voting game:

Theorem 1 (P-R Equilibrium) *Consider the voting game Γ_{PM} described above. Without loss of generality, suppose that $N_L > N_R$. Suppose further that the equation*

$$\frac{2c}{\delta}=\binom{N_L-1}{N_R}q^{N_R}(1-q)^{N_L-N_R-1}+\binom{N_L-1}{N_R-1}q^{N_R-1}(1-q)^{N_L-N_R}$$

has a solution $q^* \in [0,(N_R-1)/N_L]$.

Then the following strategy profile constitutes a Nash equilibrium:

- $\sigma_j = 1$ *for all* $j \in N_R$ *and*
- $\sigma_i = q^*$ *for all* $i \in N_L$,

Proof To prove that this profile is an equilibrium, let's analyze if for each player it prescribes an optimal strategy given the strategies of the rest. To that end, let's consider first the case of an arbitrary player $i \in N_L$. Given the strategy profile prescribed by this equilibrium, the pivot probability faced by i is:

$$\binom{N_L - 1}{N_R} q^{*N_R}(1-q)^{*N_L-N_R-1} + \binom{N_L - 1}{N_R - 1} q^{*N_R-1}(1-q)^{*N_L-N_R}$$

which, by definition, is exactly equal to $2c/\delta$. So, player i is indifferent between her two pure strategies; any randomization, including q^*, is an optimal choice.

Now we turn to the case of an arbitrary player $j \in N_R$. Given this strategy profile, j faces the pivot probability described by:

$$\binom{N_L}{N_R - 1} q^{*N_R-1}(1-q)^{*N_L-N_R+1} + \binom{N_L}{N_R} q^{*N_R}(1-q)^{*N_L-N_R}.$$

Since $q^* < (N_R - 1)/N_L$, it is straightforward to prove that this pivot probability is larger than $2c/\delta$ so that, for these players $\sigma_j = 1$ is the optimal response.

1.4 Assessing the Damage

So far we have determined that large voting games, just like the small ones, have multiple equilibria and that many of those equilibria display high levels of turnout. But the study of small games also taught us one lesson: for some cost-benefit parameters it is impossible to have an equilibrium where agents vote. In those cases it was easy to establish that whenever $c > 1/2$, the only equilibrium of the game is one where nobody votes. This turns out to be a crucial point in the study of large voting games: it determines whether or not the paradox of voting is a serious cause of concern.

To understand this, let's start by rehearsing the standard formulation of the paradox. According to the canonical view, as electorates grow larger, the pivot probability of any given player becomes smaller. At some point, that pivot probability is so small that it is no longer worth it for the player to vote, given the game's cost-benefit ratio. How seriously we should take the paradox of voting depends on how quickly these effects kick in. For the sake of argument, if we could determine that for an electorate of a 100 voters the pivot probabilities are already infinitesimal, this would be a serious problem for voting games. The key question would then be: *how quickly do pivot probabilities decline?*

But the preceding analysis shows us that it is misguided to put the question in these terms. In a voting game pivot probabilities do not have an independent existence, they are endogenous. It is not even clear what would it mean to talk about a player's pivot probability without knowing what the other players are doing. It makes no sense to ask "what is player i's pivot probability?" That is not a well-defined mathematical quantity. The correct question to ask is "what is player i's pivot probability given that the other players are choosing strategy profile σ_{-i}?" But then this means that we cannot make general statements about pivot probabilities because whatever we say will depend on the specific strategy profiles that we choose to analyze. Of all the possible strategy profiles that we could consider there are some that suggest themselves as the most relevant ones: equilibrium profiles. Thus, the key question should be rephrased as: *how quickly do* ***equilibrium*** *pivot probabilities decline?*

The short answer is "quite slowly." Slow enough to support high turnout even in very large election games with plausible cost-benefit parameters. The long answer is slightly more subtle but does not negate this basic point.

The first thing to notice is that since the pivot probabilities are endogenous, they track the cost-benefit parameter. In all the numerical examples discussed above we saw that in equilibrium players choose strategies such that the pivot probabilities faced by voters is above the cost-benefit parameter for those who use pure strategies and exactly equal to said parameter for those who randomize.

But there is an important caveat. Pivot probabilities are obtained from values of the probability density of a binomial distribution. As the total size of the electorate grows to infinity, the value of any arbitrary probability density converges to zero, although at different rates depending on the proportion of "successes" and "trials." When working with binomial distributions, it is impossible to obtain analytical expressions for the rate of this convergence. I will soon show the correct calculations when we approximate the binomial with a normal distribution. But in the meantime, let's illustrate this point with some examples.

Suppose four electorates of different sizes but with the same shares of voters N_L, N_R (55% and 45% respectively). At this point let's ignore the cost-benefit parameter and instead focus solely on the possible pivot probabilities. To that end, assume the pattern of a P-R equilibrium: every citizen in N_R votes and every citizen in N_L randomizes. What is the highest possible pivot probability? Given the properties of a binomial distribution, it makes sense to say that, whatever the value of said pivot probability, it is attained in the neighborhood of $45/55$, that is, when $\sigma_i \approx 0.818$. After all, the way to maximize the probability of exactly s successes out of t trials is by setting the probability of each individual success at s/t. In this case the answer is slightly more complicated because the pivot probability involves two different values: the would-be voter is pivotal if she breaks a tie, or forces one. But, as the following table shows, the reasoning is not far off the mark (Fig. 1.1).

This example illustrates some principles that, we will presently see, turn out to be general and have far-reaching consequences. First, the highest possible pivot probability of a voting game declines as the electorate's size grows. The reason we focus on this particular quantity is because it tells us what values of the cost-benefit parameter are compatible with a high-turnout P-R equilibrium. For example, the table shows that with 1000 citizens split 55–45, it is possible to have a P-R

No. of Players	N_L	N_R	σ_i^*	Pivot Probability
10^3	4.5×10^2	5.5×10^2	0.8188	0.0882
10^4	4.5×10^3	5.5×10^3	0.8183	0.0279
10^5	4.5×10^4	5.5×10^4	0.8182	0.00882
10^6	4.5×10^5	5.5×10^5	0.8182	0.00279

Fig. 1.1 Highest attainable pivot probabilities decline as the electorate's size grows (a numerical example)

equilibrium with approximately 900 votes so long as the cost-benefit parameter is lower than 0.0882.

Beyond this, the numerical values obtained also suggest that this decline is relatively slow. In fact, for an electorate with 100,000 citizens (100 times larger than the first one) a P-R equilibrium exists for cost-benefit parameters that are one tenth of the ones feasible with 1000 citizens. In fact, later we will see that there is a straightforward mathematical rule guiding this result. With a million voters, always with this 55–45 split, a P-R equilibrium can still exist with cost-benefit ratios of 0.002. These values, although small, are not vanishingly so. In that regard, they show that the received wisdom about the infinitesimally small pivot probabilities of voting games ought to be revised.

Now we are in a position to have a more accurate statement of the problem. As the electorate size grows, the highest feasible pivot probability, the one associated with a P-R equilibrium, declines. Therefore, the highest cost-benefit value that is compatible with said P-R equilibrium decreases. Our task is to determine how quickly this decline happens because that is what will allow us to ascertain how damaging the turnout paradox is. Thanks to the structure of the P-R equilibrium, this is a question that can be answered with substantial precision.

To that end, let's return to the equation that describes the indifference condition of the randomizing voters in a P-R equilibrium. It is:

$$\binom{N_L-1}{N_R} q^{N_R}(1-q)^{N_L-N_R-1} + \binom{N_L-1}{N_R-1} q^{N_R-1}(1-q)^{N_L-N_R} = \frac{2c}{\delta}. \tag{1.4}$$

From Stirling's formula to approximate factorials, we know that we can express any combinatorial term as:

$$\binom{n}{m} = \frac{1}{\sqrt{2\pi}}\sqrt{\frac{n}{m(n-m)}}\left(\frac{n}{m}\right)^m\left(\frac{n}{n-m}\right)^{n-m} e^R,$$

where the residual R is such that

$$0 < |R| < \frac{1}{12}\left(\frac{1}{n}+\frac{1}{m}+\frac{1}{n-m}\right).$$

Since $N_L > N_R$, let's introduce two values $0 < k_1, k_2 < 1$ such that $N_R/(N_L-1) = k_1$ and $(N_R - 1)/(N_L - 1) = k_2$. So, we can rewrite Eq. 1.4 as:

$$(2\pi(N_L-1))^{-1}\left(\frac{e^{R_1}}{\sqrt{k_1(1-k_1)}}\left(\left(\frac{q}{k_1}\right)^{k_1}\left(\frac{1-q}{1-k_1}\right)^{1-k_1}\right)^{N_L-1}\right.$$
$$\left.+\frac{e^{R_2}}{\sqrt{k_2(1-k_2)}}\left(\left(\frac{q}{k_2}\right)^{k_2}\left(\frac{1-q}{1-k_2}\right)^{1-k_2}\right)^{N_L-1}\right)=\frac{2c}{\delta}, \tag{1.5}$$

where, just like in the general Stirling's formula, the terms R_1, R_2 are residuals that converge to 0, defined so as to make the approximation exact.

The next step is to analyze the behavior of the pivot probability at a P-R equilibrium. To that end, let's set the value of q at $(N_R - 1)/N_L$. This will give us the maximal value of c compatible with high turnout. Making this substitution, the cost-benefit ratio $2c/\delta$ in Eq. 1.5 becomes:

$$\frac{(\sqrt{2\pi(N_L-1)})^{-1}}{\sqrt{k_1(1-k_1)}}\left(\left(1-\frac{k_1+1}{k_1N_L}\right)^{k_1}\left(1+\frac{k_1+1}{(1-k_1)N_L}\right)^{1-k_1}\right)^{N_L-1}+$$
$$\frac{(\sqrt{2\pi(N_L-1)})^{-1}}{\sqrt{k_2(1-k_2)}}\left(\left(1-\frac{1}{N_L}\right)^{k_2}\left(1+\frac{k_2}{(1-k_2)N_L}\right)^{1-k_2}\right)^{N_L-1}$$

The maximal cost consistent with a P-R equilibrium converges to 0 as N converges to infinity. But this equation shows us that this maximal cost converges to 0 at a remarkably slow rate. To appreciate this, after manipulating the expressions and using the fact that $(1 + 1/n)^n$ converges to e as n goes to infinity, we can see that the products of the exponential terms in this equation approach 1 as N_L becomes large. So, for large values of N, Eq. 1.5 converges to:

$$\frac{1}{\sqrt{2\pi(N_L-1)}}\left(\frac{1}{\sqrt{k_1(1-k_1)}}+\frac{1}{\sqrt{k_2(1-k_2)}}\right)=\frac{2c}{\delta}. \tag{1.6}$$

This is to say that:

$$P(V_L^i = V_R|q) + P(V_L^i = V_R - 1|q) \in O\left(\frac{1}{\sqrt{N_L}}\right).$$

Arguably, this is the central result of this chapter. Once we analyze thoroughly its implications we will see that it undermines significantly the case of the "paradox of voting" and that it buttresses the existence of high-turnout equilibria with costly voting.

In essence, what this result says is that, in the voting game we have described, costly voting is compatible with very high turnout as long as the cost remains within certain limits roughly equivalent to the inverse of the square root of the size of the winning party. This upper bound on the cost parameter is much higher than what is often assumed in the literature that does not treat the pivot probability as endogenous.

This result has another important consequence that, for reasons of space, I will not pursue further. One of the empirical regularities of voting is that electoral turnout tends to decrease with the size of the electorate (Blais and Carty 1990; Blais and Dobrzynska 1998; Geys 2006). The computations above show that the theoretical model supports this same prediction: the upper bound of the cost-benefit ratio that supports a positive participation rate declines as a function of N meaning that as the electorate's size increases, so does the level of benefits needed to induce citizens to vote.

To fix ideas about the magnitudes involved in this result, consider an electorate of 10^8 citizens, a large electorate by any standard. Suppose that 60% of this electorate favors candidate L and 40% favors candidate R. Plugging these values into Eq. 1.6 we conclude that, in this game, a P-R equilibrium can exist as long as the cost-benefit ratio c/δ is below an approximate value of 1.1×10^{-3}.

This value is much higher than what the literature on the "paradox of voting" often assumes. To interpret it correctly, we need to remember that c/δ is, in this game, the ratio between the cost of voting and the benefit a voter would obtain if she could single-handedly pick the election's winner. In this numerical example, a P-R equilibrium generates an expected turnout of 80 million voters even if the cost-benefit ratio is as high as 10^{-3}.

Is 10^{-3} too high or too low? One regretful omission in the literature on the paradox is that, for all the theoretical import of the cost-benefit ratio, there is no systematic attempt at measuring it in real-life elections. But a quick thought experiment could give us some guidance. If we suppose the cost of voting is \$1, in equilibrium a citizen would find it rational to vote if, from her point of view, the benefit from deciding the election with her vote is \$1000. Of course, if the benefit is larger than \$1000 (or, conversely, if the cost is smaller) she will also vote.

I submit that such cost-benefit ratio is highly plausible and that, if anything, most voters would consider it too high. If my assessment is correct, then the game-theoretic analysis of voting predicts high turnout even in very large electorates.[2] In other words, in the voting game that we have been analyzing, for a highly plausible set of parameters the conclusion is clear: there is no such "paradox of voting."

[2]To put this point more starkly, with an electorate of 6×10^9 (roughly the entire world's population), if it were evenly split, a cost-benefit ratio of 1.5×10^{-5} would support a P-R equilibrium with almost everyone voting.

1.4.1 A Fateful Detour

Much of the results just discussed have been known since Palfrey and Rosenthal's paper of 1983. (Although Palfrey and Rosenthal did not carry out the calculations about the rate of decline I just presented.) They show that large elections are perfectly compatible with high levels of turnout. And yet the "turnout paradox" is still regarded as a major theoretical roadblock. Why? The reason is that, in a laudable display of intellectual honesty, Palfrey and Rosenthal (1985) set out to prove that the high-turnout equilibria of their previous contribution were, in their words, "fragile." Understandably, such rejection carried enormous weight; not only the critics were the very authors of the result, they were (already by then) some of the most accomplished scholars in the field. But their criticism was misguided, as I will now show.

Palfrey and Rosenthal base their attack on the high-turnout equilibria they had already obtained on two arguments. First, they claim that these equilibria are implausible, "very fragile" (Palfrey and Rosenthal 1985, p. 65), because they require very exact partitions within the groups of voters and so, are unlikely to materialize. Second, they claim that those equilibria do not survive in a generalized version of the model where voters have imperfect information about each other's payoffs. Let's take these arguments one at a time.

1.4.2 Balancing Groups

The first claim is not clear because the authors do not offer a precise definition of what makes an equilibrium fragile. The closest they come to offering a clear definition of the problem is the following passage:

> Although this result is very interesting because it suggests that substantial voting can occur when expected utility maximizing voters have relatively high voting costs and there is no citizen duty, the type of equilibrium that supports high turnout seems very fragile. In particular one side must be split into subgroups of voters and nonvoters in a precise way so that there is no uncertainty about how many votes one of the two alternatives will receive, because if there is too much uncertainty in equilibrium, then the probability that the election is close (i.e., either a tie or one vote away from a tie) approaches 0 as N gets large. (Palfrey and Rosenthal 1985, p. 65)

There are several problems with this argument. First, it is not possible to determine if an equilibrium is fragile or not simply by looking at its description. To determine if an equilibrium is fragile it is irrelevant whether it is odd-looking or not. It is telling that in standard economic analysis this argument would never be made. For instance, if we believe that a price p equilibrates demand and supply in the market for wheat, we are implicitly saying that at this price some would-be producers will not grow wheat at that price, whereas others will and that some would-be consumers will not buy wheat at that price whereas others will. To say that the market of wheat is in equilibrium at price p is to say that the participants in that market will split in a precise way. If we were to apply the standards adopted by Palfrey and Rosenthal to this case we would have to discard this equilibrium on the grounds that it is exceedingly "precise." Yet no economist would ever consider this possibility.

In the standard analysis of equilibrium, say in microeconomics, equilibrium E is fragile (or unstable) if, whenever the system is in it and is subjected to a tiny shock, that shock sets in motion a process that leads the system farther away from E. In contrast, E would be robust (or stable) if, in response to said shock, there are countervailing forces that bring the system back to E. No matter how contorted an equilibrium seems at first glance, if there are overwhelming forces that push the system in that direction, we must concede that it is, in fact, robust. Later, in Chap. 3 I will use the method of stability sets to carry out the type of analysis that we need to make precise this idea. The results there will show that these equilibria with high-turnout are quite robust indeed.

As a matter of fact, in their analysis Palfrey and Rosenthal end up singling out, among all the game's equilibria, the one where all players randomize, that is, the mixed-strategy equilibrium. This equilibrium, as they have shown repeatedly, is such that as the number of citizens grows, electoral participation decreases, just as the "paradox of voting" would claim. The problem is that, as our subsequent stability analysis will show, mixed-strategy equilibria are the most fragile ones. Their stability set is of measure zero, meaning that a game arrives at its mixed-strategy equilibrium with probability zero.

We could also interpret this remark as claiming that the specific type of behavior needed for a P-R equilibrium is so tightly specified that it would be nothing short of a miracle to observe it in any application. But this overlooks the enormous multiplicity of equilibria of the game. The numerical example just presented shows the problem with this argument. Palfrey and Rosenthal seem disturbed by the fact that the P-R equilibrium requires players to split in a very specific way (all citizens in N_L randomizing, all citizens in N_R voting). But, what if they do not split this way? They might well split in a different way that may also be an equilibrium. For example, we just saw that a split where all of N_R randomize, six million citizens in N_L abstain and four million vote is also an equilibrium. If it seems that this could only happen as a result of some very felicitous accident, that is not a problem either because slightly different splits are also equilibria. For example, we just saw that a split of 3,999,999 N_L voters and 6,000,001 N_L abstainers is also an equilibrium or that, for that matter, another split with 3,999,000 N_L voters and 6,001,000 N_L abstainers is yet another equilibrium. In short, this definition of fragility is simply misguided and there is no reason to use it as a basis for discarding the high-turnout equilibria of a voting game.

Equilibria like the ones I just showed may be criticized by virtue of being asymmetric and, although they do not state it in so many words, Palfrey and Rosenthal seem to be uneasy about them precisely on that count. In fact, the equilibria with 13 million voters in the previous example require members of N_R with the same cost-benefit parameter to choose different strategies. This concern is misplaced. Arguably, there might be something wrong about asymmetric equilibria in symmetric games (although even that can be contested). But there is nothing objectionable about asymmetric equilibria in *asymmetric* games. So far we have analyzed games where all the players in each group have identical payoffs. But for the purposes of a general theory of voting, these are an utterly uninteresting case. What we really want to analyze is asymmetric games where different players have different payoff functions. So, if we want to deploy this objection, we need to show that it is not possible to support high turnout equilibria in asymmetric games. As we will soon see, asymmetric games do not pose any problem in this regard.

1.4.3 Imperfect Information

The second and more important sense in which Palfrey and Rosenthal call these equilibria fragile is that they do not survive once we relax the assumption of perfect information. If true, this charge is damning because any relevant application of voting games should deal with the fact that in real life information is imperfect. To prove it, the authors develop an imperfect-information version of the voting game and, with correct judgment, determine that the relevant solution concept is the Bayesian Nash equilibrium. Therefore, they introduce uncertainty about the cost parameter. So, for the purposes of this analysis, c is no longer a constant but a random variable with a distribution governed by the distribution functions $F_L(c), F_R(c)$ where the subindex refers to the relevant group of voters. For the time being, we will keep considering δ as a constant. Not only is this in keeping with Palfrey and Rosenthal's analysis (they normalize δ at 1), it does not affect the results at all. Here, however, they make a subtle but important mistake: when imposing some structure on this distribution function, they assume (Assumption 2.ii., p. 70) that $F_L(1) < 1, F_R(1) < 1$.

Intuitively, this means that every voter considers that with some positive probability, the cost of voting for any other voter is larger than 1 (or at least equal to 1). The authors offer two defenses for this assumption and I will discuss them in the order in which they appear. Ultimately I will conclude that none of these defenses is satisfactory and that this assumption ought to be abandoned.

The first defense claims that "(this assumption) is needed to guarantee that in large electorates essentially the only voters are citizen-duty voters (i.e., $c_i \leq 0$)." (Palfrey and Rosenthal 1985, p. 75). This is a weak defense. In essence it says that the assumption is justified because it establishes the result they want to prove, what they call the Downs-Riker-Ordeshook-Tullock view. But the whole point of their masterful game-theoretic analysis was to assess whether or not rational citizens with a positive cost of voting would vote. Their earlier results had answered this question in the affirmative, ultimately refuting the citizen-duty theory. Instead, in this attempt at generalizing their analysis, they introduce an assumption that stacks the deck in favor of such theory without offering any good reason for doing so. There is no need to guarantee that the citizen-duty approach is correct. Either it is or it is not and the right way to decide on this matter is, like they had done before, to develop a model that does not assume such result and take its analysis to its last consequences whatever they are. If, as happened in their 1983 piece, the citizen-duty theory is proven wrong, so be it.

The second defense they offer is that they "feel it is a realistic assumption that *some* voters find it too costly to vote, regardless of the probability that they would be decisive" (Palfrey and Rosenthal 1985, p. 77, my emphasis). This defense expresses verbally a plausible condition they wish to impose on the model: that, with certainty, there are some voters that consider voting as too costly to be worth it. But the mathematical assumption imposes a much stronger and implausible condition. It says that, with some probability, every voter considers voting too costly to be worth it. This is the key mistake. Although it is reasonable to want to build into the model some voters that will always find it too costly to vote, Assumption 2.ii. does much more than this. As we will see, it ensures that *most* citizens do not vote. That can hardly be a reasonable assumption; it is not what Palfrey and Rosenthal intended to assume. But, because their analysis overlooked the exact behavior of the pivot probabilities, this is what they unwittingly did.

The analysis we have carried out above allows us to calculate with some precision how many citizens will not vote under any circumstance given Assumption 2.ii. Let's look at the example we have already studied to illustrate this. We determined that, with 100 million citizens, as long as the cost-benefit ratio remains below 10^{-3} it is still possible to sustain a P-R equilibrium provided that the partition between parties is not exceedingly lopsided (in this example, a 60–40 partition was still acceptable). It is worth recalling that this cost-benefit ratio is quite reasonable. At least, there is no a priori reason to reject it.

Now let's introduce Assumption 2.ii so that both groups of citizens (L and R) face a uniform distribution $U[0, 1]$ on their cost-benefit parameter. The first thing to notice is that under this assumption, half of the electorate faces a cost-benefit ratio $c > 1/2$. From the study of small voting games we know that if a player's cost-benefit ratio is more than 1/2, she will not vote regardless of the size of the electorate. Not even in a three-player game will she vote because, twice this cost-benefit ratio will always be larger than any pivot probability, which is by definition, less than 1.

This should already alert us that there is something inadequate about Assumption 2.ii. It stipulates that half of the electorate will not vote under any circumstance, even in very small elections because, for them, the benefit of deciding the election is not even twice the cost of voting: if for them voting costs \$1, they would not be willing to pay \$2 for the possibility of deciding the election. Such ratios would give new meaning to the words "voters' apathy."

This problem is compounded once we look at large elections. In this example, under this assumption 99.9% of the electorate have a cost-benefit ratio higher than 10^{-3} which, as we already know, is what is needed to secure positive turnout. But 10^{-3} is already a pretty high ratio, arguably much higher than is realistic. The "some voters" to which Palfrey and Rosenthal refer in their stated defense of Assumption 2.ii. should rather be taken to mean "almost all voters." Under such conditions, it is not surprising that the authors have a hard time preserving acceptable levels of turnout.

At first glance, Assumption 2.ii. seems an innocent normalization that should have no effect whatsoever on the final results and would simply allow for a small amount of players that do not vote at all. But once we place it in the context of what we know about the behavior of pivot probabilities in voting games, it turns out that it depresses turnout without any real justification.

It is a testament to these authors' technical prowess that even their mistakes are brilliantly executed. The preceding discussion raises the question of what happens under an alternative setting. Palfrey and Rosenthal have worked out the answer to this question in that same paper. In the paper's second appendix, in the very last page (p. 77), they consider a model in which $F(3/8) = 1$. This places a cap of 3/8 on the value of c. The result is that the game now has multiple equilibria, some of them with high levels of turnout even as N grows to infinity. It may seem at first glance that this model cannot capture the fact that some voters may find the cost of voting too high to be worth it. But we could easily introduce a group of such citizens here, possibly setting their cost $1 < c < 2$ so that, they will not vote regardless of their probability of being pivotal. The presence of those citizens would not affect the equilibrium strategies of the rest because they are immaterial from the point of view of the election's outcome. Of course, these citizens will dilute the turnout rate but, unless we let their number converge to infinity, they will not drown out turnout entirely. For example, if we have high turnout equilibrium with voting rates of 77.76% (just as before) and we add a number of such citizens equal to 20% of the total citizenry, turnout will drop to 64.8%. As a result, this would be a model with imperfect information, majority rule, citizens with high costs of voting, in a word, all the desiderata that Palfrey and Rosenthal impose, and yet it would have a high turnout equilibrium.

If we do not stack the deck so heavily against the P-R equilibria, it is possible to construct games with imperfect information with high turnout. Following Palfrey and Rosenthal's analysis of this model, let's introduce two additional items of notation so that $\gamma_L(c)$ represents the number of citizens in N_L such that their cost-benefit parameter $c_i < c$. Likewise, $\gamma_R(c)$ will stand for the equivalent number of citizens in N_R. Just as with vote totals so far, γ^i will represent the value of γ with player i removed. Palfrey and Rosenthal have proven that, in this setting, we can characterize the game's Bayesian Nash equilibrium with a pair of critical cost levels c_L^*, c_R^* such that any player in N_L votes if her cost parameter is lower than c_L^* and abstains otherwise, with an exactly analogous condition for voters in N_R. These critical levels of this cutoff-point equilibrium are determined by the following conditions:

$$\frac{2c_L^*}{\delta} = P(\gamma_L^i(c_L^*) = \gamma_R(c_R^*)) + P(\gamma_L^i(c_L^*) = \gamma_R(c_R^*) - 1), \tag{1.7}$$

$$\frac{2c_R^*}{\delta} = P(\gamma_R^j(c_R^*) = \gamma_L(c_L^*)) + P(\gamma_R^j(c_R^*) = \gamma_L(c_L^*) - 1). \tag{1.8}$$

To verify that the decision rules embodied by these critical levels constitute an equilibrium, let's consider the payoff of an arbitrary i player for either of her strategies. If she votes, her expected payoff is:

$$(u_L(x_L) - u_L(x_R))(P(\gamma_L^i(c_L^*) \geq \gamma_R(c_R^*))+$$
$$1/2P(\gamma_L^i(c_L^*) = \gamma_R(c_R^*) - 1) + u_L(x_R) - c_i.$$

If she abstains, the expected payoff is:

$$(u_L(x_L) - u_L(x_R))(P(\gamma_L^i(c_L^*) > \gamma_R(c_R^*))+$$
$$1/2P(\gamma_L^i(c_L^*) = \gamma_R(c_R^*) - 1) + u_L(x_R).$$

So, her expected value of voting will exceed that of abstaining if and only if $c_i > c_L^*$. The same reasoning holds for an arbitrary $j \in N_R$. This analysis is analogous to the one we conducted above while computing the best-response so it is not surprising that we arrive at a similar answer. This should already alert us to the fact that the findings in this game will not differ widely from the ones we made in the case of perfect information.

In fact, if, instead of introducing the implausible Assumption 2.ii. we simply stipulate that the cost parameter is a random variable with a distribution defined over the interval

$$[0, \sqrt{2\pi(N_L-1)}^{-1}(\sqrt{k_1(1-k_1)}^{-1} + \sqrt{k_2(1-k_2)}^{-1})],$$

we are back to the previous case of a game with a P-R equilibrium in high turnout.

To see this, let's go back to the numerical example we discussed above. In this case, let c/δ follow a uniform distribution $U[0, 10^{-3}]$. Then, there exists a Bayesian Nash equilibrium where $(c_L/\delta)^* \approx 2/3 \times 10^{-3}$ and $(c_R/\delta)^* = 10^{-3}$. That is, in this equilibrium every citizen in N_R votes and roughly two-thirds of the citizens in N_L vote. It is easy to verify that these decision rules satisfy the equilibrium conditions in Eqs. 1.7 and 1.8. Just as in the case above, this equilibrium results in a turnout of around 80 million voters.

But, it may be argued, it is necessary to introduce in the model some voters that will not vote, no matter how decisive their vote is, which is what Assumption 2.ii accomplishes. It remains an open question if such voters do, in fact, exist. We all know people who seem pretty indifferent to political matters and, as a result, never bother to vote. But we never have the chance to observe how they would behave in an election if they were absolutely certain that their vote is decisive.

Even if we wanted to include such types of voters in our model, there are better ways of doing it than with the extremely large parameters of Assumption 2.ii. For instance, we could assume in this numerical example that c/δ is distributed in the interval $[0, 1.2 \times 10^{-3}]$. This would not change the value of the critical points and would merely result in a turnout of roughly 60% of the electorate, still pretty high, as a result of the fact that now some 20% of the citizens are so indifferent between the candidates that they would not vote even if they knew their vote would decide the election.

In sum, the notion that high-turnout equilibria disappear once we introduce imperfect information is highly overstated. To be sure, imperfect information can destroy a P-R equilibrium, but only if we assume cost parameters far above what is warranted. After all, fragility and robustness are relative concepts. Relative, that is, with respect to the size of the shock administered. Once we know the behavior of the endogenous pivot probabilities, it becomes clear that Assumption 2.ii. administers to the model a cataclysmic shock of implausible dimensions that hardly any positive-turnout equilibrium can be expected to withstand. If we want to pronounce the P-R equilibria fragile or robust, we must test them with perturbations within the limits of the model's pivot probabilities. As it turns out, the P-R equilibria pass those tests without a problem: even after introducing imperfect information, duly parametrized voting games are not afflicted by the paradox of voting.

1.5 Heterogenous Voters

In spite of its problems, the analysis that Palfrey and Rosenthal offer of voting games with incomplete information has an important by-product: it serves as the basis to generalize the pure-majority game to the case of voters with heterogenous payoffs. Voting models where all the sympathizers of one party have the same preference are only of mathematical interest. Most of the fundamental questions that a theory of voting should address pertain to voter heterogeneity. For example, if we want to assess the functioning of representative democracy, we need to be able to ascertain if and how parties respond differently to extremist and centrist voters. This is all the more so once we open our analysis to the problems of turnout. With full participation, the only way in which a voter can defect a party is to vote for the opponent. But with variable turnout citizens can simply sit the election out.

As it happens, in light of the results obtained by Palfrey and Rosenthal, the case of heterogenous voters is quite simple. In fact, the Bayesian Nash equilibria that they describe for a game with imperfect information and homogenous voters are at the same time the Nash equilibria of a game with perfect information and heterogenous voters. I will not enter into the details here, but a quick look at the definition of both equilibria shows why this is the case.

A Nash equilibrium is a strategy profile where every player's strategy is optimal given the choices of the remaining players. In a Bayesian Nash equilibrium, a player's strategy is optimal given his beliefs about the distribution of the other players' types and the actions that each type will choose. In the game analyzed by Palfrey and Rosenthal a type is identified by the payoff function, more exactly, by the cost-benefit ratio. Since in this game there are no private signals, the posterior beliefs that each player holds about the distribution of types of each player coincide with the common prior. In other words, a Bayesian Nash equilibrium gives us, for every type of player, that is, for every cost-benefit ratio, the optimal strategy given the entire distribution of the cost-benefit ratios. Intuitively it is clear that this corresponds to the Nash equilibrium of a game with heterogenous voters and perfect information. I will not offer a formal proof of this statement but, as we will see in the next sections of this chapter, the cut-off point equilibria described by Palfrey and Rosenthal also serve as Nash equilibria for the more general model of heterogeneity in the cost-benefit ratio. It may be asked if we should not generalize the model to cover the case of heterogenous voters and imperfect information. But it is far from clear what would be gained from this. In these early stages of developing a theory of aggregate voting behavior we do not need to concern ourselves with private signals. Undeniably, information plays an important role in any election but, at least as a first approximation, the information that matters is public. Thus, the posterior beliefs of any individual player will be equal to the prior beliefs: the description of the equilibrium based on said priors will then suffice.

1.6 The Problem with the Margin of Victory

After defending the P-R equilibria from the main objections it has faced, I now want to formulate an entirely different objection. The most attractive property of the P-R equilibria is that they generate high levels of turnout, even for relatively high costs of voting and that, as we have just seen, such levels are robust to the introduction of imperfect information. This is accomplished because in a P-R equilibrium the pivot probability is endogenous. But then the high turnout comes at a price: the model generates implausibly high pivot probabilities or, which comes to the same, implausibly narrow margins of victory. A numerical example will be useful. For consistency, let's use again the same numerical parameters than before but, so as to stack the deck in favor of the P-R equilibrium, let's consider a cost parameter of 10^{-5}. That is, $N_L = 6 \times 10^7$ and $N_R = 4 \times 10^7$. In a P-R equilibrium, turnout is, again, approximately 80 million voters, with all citizens in N_R voting and all citizens in N_L randomizing at rates close to 2/3 so as to make the pivot probability of each voter equal to 5×10^{-6}.

With these parameters roughly 99.3% of the times the margin of victory will be less than 10,000, an advantage of 0.0125%. In this model, winning an election by 0.5% would be considered a landslide. (I should add that Palfrey and Rosenthal had already made a similar observation only that they did not regard it as pathological.)

Clearly, these margins of victory are not realistic. This problem is the flip side of the high participation rates of P-R equilibria. In fact, since in these equilibria the pivot probability endogenously matches the cost of voting, with large numbers of voters this can only happen if the two groups vote at very similar rates. It seems, then, that although we do not have a paradox of voting "we do have a paradox of surplus voting." That is, the game-theoretic model of voting can explain high levels of turnout, but seems unable to explain large margins of victory. If in the preceding sections I defended the canonical model, now I will argue that, given this problem, this time around we really need to amend it but, and this is crucial, without rejecting it.

Chapter 2
A General Model of Strategic Voting

2.1 Introduction

In the last chapter we studied the canonical model of voting and came to the conclusion that the "turnout paradox" is not really a cause of concern in the game-theoretic study of elections because it is easy to generate high-turnout equilibria for very plausible cost parameters. Some relatively minor problems remain. In particular, this model cannot account for realistic margins of victory. Furthermore, its equilibria tend to be hard to describe since they require combinations of pure and mixed strategies for large numbers of players. In this chapter we will see that these difficulties are the result of some excessively restrictive assumptions of the model and that, once we consider more general, and more plausible specifications, they are easily overcome.

2.2 Generalizing the Outcome Function

The standard game-theoretic model of voting assumes a step-wise outcome function so that a candidate obtains absolute victory with one more vote than the opponent or, conversely, goes down to total defeat with one vote less. So, from the standpoint of the voters, in computing the best-response correspondences, the only events that matter are a tie or a victory by one vote. Given that structure, it is understandable that in equilibrium the voters' strategies are such that these two events are very likely, at the expense of other more comfortable margins of victory. But in real life, the assumption that only victory (by one vote) is what matters is clearly unrealistic.

Consider, for instance, the fate of losing parties in an election. The model we have been studying assumes that the loser will have absolutely no influence in the policy choice, regardless of his vote. There are many reasons why this is wrong.

© Springer International Publishing AG 2018
L. F. Medina Sierra, *Beyond the Turnout Paradox*, SpringerBriefs in Political Science,
https://doi.org/10.1007/978-3-319-73948-9_2

First, as anyone familiar with the story of the US presidential election in Florida in 2000 knows, the final tally is not a deterministic function of the vote. Human and machine mistakes intervene, not to mention, in some countries, fraud. If a candidate obtains five votes less than the other, a few glitches in the system may put him on top. Instead, if defeated by several millions of votes, only a massive fraud of the kind we do not need to consider when modeling a democracy may bring him to victory.

Furthermore, a democracy requires a sequence of elections, not just one. So, an election determines also the long-run viability of a party. A party that obtains 38% of the vote is a force to be reckoned with in future elections. It will be able to benefit from financial contributions and state-provided funds, its organization, although shaken, is likely to remain in place. As such, it will have a say in future policy-making. Instead, a party that obtains 0.02% of the vote can be safely ignored, not only by the government but also by contributors and all but the most die-hard activists. If it survives to contest the next election, it will be solely on the strength of, and for the benefit of, its handful of loyalists.

Just as future elections, past elections also matter. The victor in a democracy does not make policy from scratch but has to deal with the legacies of the past. Part of the public administration will be holdovers from previous governments, some of them ideologically hostile to it, and this will hamper its ability to shape policy to its liking. It will also have to deal with courts whose composition may reflect the views of previous governments. In sum, winning an election does not immediately translate into obtaining the ideal policy. There are many hurdles to overcome, a task made easier with a high margin of victory.

Constitutional rules also ensure that real-life political processes do not resemble the one represented by the canonical model. Even in the Westminster system of complete majority rule, that is, with no proportional representation, the minority MPs are not simply fired from Parliament. This means that the majority's margin for action is larger, the larger it is. A majority of one gives substantial bargaining power to a handful of members of the coalition, usually those who are ideologically closer to the minority.

All these considerations indicate that the step-wise function of our pure-majority model is an analytical straitjacket. To be sure, it makes the computation of some equilibria easier but even in this regard the gain is not substantial: as we have seen, even the simplest P-R equilibrium needs to be computed numerically. Beyond that, they also suggest the direction in which we might take the analysis to make it more general.

In principle it would be desirable to find an outcome function that captures all the relevant institutional details of the specific polity under study. But that is beyond the purview of this book, not the least because my goal here is to develop the most general results possible, covering the largest possible scope of cases.

The following function, which I will call the "generalized outcome function," has some very useful properties that I will discuss presently. If, just as before, we define $V_L|s$ and $V_R|s$ as the number of votes for x_L and x_R respectively, for a given strategy profile s and we introduce a parameter $\rho > 0$, we can redefine the outcome function as:

$$g(V_L, V_R|s) = \frac{x_L e^{\rho(V_L - V_R|s)}}{1 + e^{\rho(V_L - V_R|s)}} + \frac{x_R}{1 + e^{\rho(V_L - V_R|s)}}. \qquad (2.1)$$

This function is easily recognizable as a variant of the logistic function. It is increasing in L's margin of victory and increases smoothly between x_R and x_L as this margin increases. Such smoothness captures the intuition of the preceding discussion: no single vote is responsible for absolute victory or defeat. Furthermore, it has the advantage of being very flexible, a good attribute given the level of generality at which the present analysis is conducted. So, the parameter ρ can be interpreted as representing the degree of proportionality in the underlying electoral system. If $\rho = 0$, the outcome is just the average of x_L and x_R regardless of each policy's vote share. This would be the equivalent to a Platonic consociational regime where power-sharing is so strict that the minority always gets half of everything, no matter how small its vote. As ρ goes to infinity, the outcome function behaves more and more as the original step-wise function.[1]

For our purposes, this outcome function has the advantage of been relatively simple to compute. Other functions could have been chosen, conceivably with more realistic properties. But the main results, that is, the properties about turnout, would have not changed. Since the goal of the present analysis is illustrative and theoretical, not computational and empirical, the present function seems an acceptable choice.

In what follows I will adopt the more general framework of heterogenous players. That is, I will assume that the cost-benefit parameter differs across citizens so that, instead of being one common value c for all of them, each voter i will have an individual parameter c_i. Once we have the solution for this game, it is easy to retrieve the solution for the simple case with homogenous players although, at this stage, it is hard to see what purpose that would serve.

Although in this analysis heterogenous voters means simply voters with different cost parameters, it could also mean voters with different benefits. The analysis would remain the same. This is something that resonates with the broader goal we set out at the beginning: a model of turnout with heterogenous voters is a model of turnout where voters' preferences over the candidates differ, just as we would expect from a spatial model of electoral competition.

---

[1] Castanheira (2003) has analyzed a similar outcome function obtaining similar results. There are, however, differences that make the result in that paper and the one here complements rather than substitutes. Castanheira's model assumes population uncertainty, following the idea introduced by Myerson (1991) and assumes from the outset that the policy function is continuous or equivalently, which is the formulation he prefers, that voters care about the margin of victory. We have seen here that high-turnout equilibria exist even in the canonical case of a fixed population. In that sense it would be incorrect to claim that it is only thanks to Poisson models or continuous outcome functions that we can "solve" the voting paradox. As regards the rationale behind the function, Castanheira considers it as a case in which voters care not just about the victor of the election, but about the margin of victory. Although an undoubtedly plausible argument, I believe it introduces a rather unnecessary change in the players' payoff function. As I have already made clear, I prefer to interpret the continuous outcome function as reflecting the fact that the policy-making process itself is such that its outcomes respond continuously to the margin of victory.

Now the outcome is a linear combination between the two options. So, to evaluate the voters' payoffs in this case we need to impose an extra assumption on the Bernoulli utility functions. The simplest such assumption is, of course, linearity. Since in our subsequent analysis we will not be interested in the effect of the voters' attitudes toward risk, then there is no harm in making such assumption. Then, we can calculate the expected payoff of each strategy. In the case of an arbitrary agent i in N_L these are:

$$w_i(V_i, s_{-i}) = u_L(x_L)g(V_L^i + 1, V_R) + u_L(x_R)(1 - g(V_L^i + 1, V_R)) - c_i$$

$$= \frac{\delta e^{\rho(V_L^i + 1 - V_R|s_{-i})}}{1 + e^{\rho(V_L^i + 1 - V_R|s_{-i})}} + u_L(x_R) - c_i \tag{2.2}$$

$$w_i(A_i, s_{-i}) = u_L(x_L)g(V_L^i, V_R) + u_L(x_R)(1 - g(V_L^i, V_R)) \tag{2.3}$$

$$= \frac{\delta e^{\rho(V_L^i - V_R|s_{-i})}}{1 + e^{\rho(V_L^i - V_R|s_{-i})}} + u_L(x_R), \tag{2.4}$$

By comparing these two expressions, we obtain the following best-response correspondence:

$$\sigma_i^*(s_{-i}) = \begin{cases} 1 & \text{if } g(V_L^i + 1, V_R|s_{-i}) - g(V_L^i, V_R|s_{-i}) > \frac{c_i}{\delta}, \\ [0,1] & \text{if } g(V_L^i + 1, V_R|s_{-i}) - g(V_L^i, V_R|s_{-i}) = \frac{c_i}{\delta}, \\ 0 & \text{if } g(V_L^i + 1, V_R|s_{-i}) - g(V_L^i, V_R|s_{-i}) < \frac{c_i}{\delta}. \end{cases}$$

As usual, an analogous expression holds for citizens in N_R. Since these best-response correspondences are very similar to the ones we analyzed before, it is a good idea to focus on the equilibria that can be described in terms of critical values of the cost-benefit parameter.

Theorem 2 *Let Γ_{GO} be a voting game with heterogenous voters and outcome function defined by Eq. 2.1, and let the critical values $(c_L/\delta)^*$ and $(c_R/\delta)^*$ be defined implicitly by the functions:*

$$\frac{e^{\rho(\gamma_L(c_L^*) - \gamma_R(c_R^*) + 1)}}{1 + e^{\rho(\gamma_L(c_L^*) - \gamma_R(c_R^*) + 1)}} - \frac{e^{\rho(\gamma_L(c_L^*) - \gamma_R(c_R^*))}}{1 + e^{\rho(\gamma_L(c_L^*) - \gamma_R(c_R^*))}} = \frac{c_L}{\delta}^* \tag{2.5}$$

$$\frac{1}{1 + e^{\rho(\gamma_L(c_L^*) - \gamma_R(c_R^*) - 1)}} - \frac{1}{1 + e^{\rho(\gamma_L(c_L^*) - \gamma_R(c_R^*))}} = \frac{c_R}{\delta}^* \tag{2.6}$$

Then, a strategy profile σ^* is a Nash equilibrium of Γ_{GO} if, whenever $c_i < c_L^*$, $\sigma_i^* = 1$ and whenever $c_i > c_L^*$, $\sigma_i^* = 0$. (An analogous statement holds for c_j.)

Proof Consider a voter $i \in N_L$ such that $c_i > c_L^*$. Duly modified, the argument here will work for $c_i < c_L^*$ and for every $j \in N_R$. At the equilibrium profile σ^*, i's payoff from voting is $g(V_L + 1, V_R|s^*) - c_i$ and the payoff from abstaining is $g(V_L, V_R|s^*)$. Rearranging terms, from Eq. 2.5 we conclude that $w_i(A_i, \sigma_{-i}^*) > w_i(V_i, \sigma_{-i}^*)$ so that in equilibrium, i's optimal strategy is A_i.

This equilibrium is not subject to the "paradox of voting" any more than the previous P-R equilibria. To illustrate this, consider a numerical example with the same parameters as before, that is, $N_L = 6 \times 10^7$ and $N_R = 4 \times 10^7$. For simplicity's sake, suppose that the individual cost-benefit ratios are distributed in such a way that, for every $j \in N_R$, $c_j/\delta \leq 10^{-5}$ and for $i \in N_L$, 66.6675% of voters have a value $c_i/\delta \leq 10^{-5}$. Furthermore, suppose that in the outcome function $\rho = 4 \times 10^{-5}$. Then, there is an equilibrium with cutoff points $(c_L/\delta)^* = (c_R/\delta)^* = 10^{-5}$ where every citizen in N_R votes as well as 40,000,500 citizens in N_L. This is a turnout of roughly 80 million voters.

But, beyond its high turnout, which at this point should no longer constitute a surprise, this example illustrates one important advantage of the generalized outcome function with respect to the step-wise version adopted before. Under the new function, the voting game's equilibria are such that, except for a particular cost level, every player chooses a pure strategy. Just like in a tipping game, there is a critical level, in this case one for each team, below which every agent participates and above which none does. For analytical purposes this means that it is possible to compute the equilibria as the fixed point of two relatively simple equations, without dealing with awkward randomizations.

To understand why this is the case we need to understand why there are no pure-strategy equilibria in the pure-majority game. By looking at that game's best-response correspondence (Eq. 1.3), we see that the only players in N_L that choose a pure strategy are those for whom the value $2c/\delta_i$, that is, their cost-benefit ratio, is above or below $P(T_L^i = T_R|\sigma_{-i}) + P(T_L^i = T_R - 1|\sigma_{-i})$ (an analogous expression holds for players in N_R). This latter term, as we know is the probability of being pivotal for any given voter. But if every player except i is choosing a pure-strategy, this term is either 0 or 1, which means that, in a Nash equilibrium, player i would only choose a pure strategy if her term $2c/\delta_i$ is either larger than 1 or less than 0, which is to say, player i will vote only if $c < 0$ or will abstain only if $c > \delta_i/2$. The first case ($c < 0$) is equivalent to the "civic duty" voters already discussed and the second case ($c > \delta_i/2$) represents voters for whom the cost of voting is unreasonably high, higher than twice the benefit of single-handedly deciding the election. For all the cost parameters in between these two extremes, which is to say, for virtually all the parameters of analytical interest, if every other voter is choosing a pure strategy, the voter in question does not have an equilibrium response in pure strategies. By contrast, if and only if, at least some voters choose mixed strategies, then the pivot probability takes values between 0 and 1, thus making possible the choice of equilibrium strategies for other voters.

As we can see from the best-response correspondence of the game with a continuous outcome function (Eq. 2.2), in this framework it is no longer the case that the critical value generated by a pure-strategy profile is either 0 or 1. This means that the existence of a pure-strategy equilibrium is no longer subjected to the unreasonable restrictions on the cost-benefit parameter that are necessary in the pure-majority case.

This new model will serve as the bedrock for the subsequent results. Later (Sect. 2.4), I will show how to use this model to compute equilibria of voting games that will help us to obtain comparative statics results as a function of the structural parameters, the ultimate goal of this entire exercise. But before going into this, I shall study some variants of the model that will illustrate its flexibility. Some of these variants will also overcome the main difficulty we encountered in the pure-majority case: the excessively narrow margins of victory.

2.3 Some Extensions and Variations

2.3.1 Quantal-Response Equilibrium

Beginning with the work of Palfrey and McKelvey (1995), in recent years we have seen an interesting generalization of the standard concept of Nash equilibrium: quantal-response equilibrium (QRE). It is a concept that builds on an old intuition but that corrects several common misunderstandings around it, offering instead a rigorous and fruitful formulation. The inherent unpredictability of human decisions has been an old topic and cause of concern among game theorists. A long-standing tradition has tied this with the role of mixed strategies in games, suggesting that such strategies are the ones that best capture the fact that human actions are never completely deterministic.[2] Reasons of space prevent me from discussing why I believe that view is wrong. Suffice it to say that we ought to distinguish between the uncertainty resulting from the inherent randomness of individual choices and the uncertainty resulting from the strategic choices of fully rational actors, as is the case in mixed-strategy equilibria. The reader interested in knowing my own views on mixed-strategies can consult my *A Unified Theory of Collective Action and Social Change* (Medina 2007). One way to avoid this confusion is to start, from the outset, assuming that individuals can make random mistakes in their choices and then studying what equilibrium behavior would imply in that case. This is what QRE is meant to accomplish.

In a recent paper Levine and Palfrey (2007) apply the notion of QRE to voting games and, after subjecting it to a very instructive experimental test, conclude that it is possible to generate high turnout equilibria in this framework so that, in their words, "QRE effectively resolves the paradox" (p. 155). It is an outstanding result so that some comments are in order.

As the analysis in the preceding chapter makes clear, I take exception to the concluding claim. We already saw that, if we rule out uninteresting parameterizations, it is possible to generate high-turnout Nash equilibria in voting games with costly voting. In the case of Palfrey and Levine, they only compare QRE with the mixed-strategy Nash equilibrium which, as is well known, is characterized by asymptotically vanishing turnout, unlike the equilibria that use pure strategies. As I already discussed above, these pure-mixed equilibria are not fragile and, hence,

---

[2]A fascinating discussion of this topic can be found in the exchange between Osborne and Rubinstein (1994) in their jointly authored *A Course in Game Theory*.

should be part of an integral analysis of voting games. But this is a minor point. What matters is that, thanks to the application of QRE to voting games we can move a step closer to an understanding of the general properties of these games. In fact there is a revealing connection between the QRE analysis and the model I presented in the previous section.

At first glance it may seem that there is no connection whatsoever. After all, QRE is an equilibrium concept developed to cover the case where individual players can make mistakes with some positive probability, while I analyzed the preceding model using the deterministic responses typical of Nash equilibrium. But, mathematically, both models share a common property: the function used to determine each player's best-response correspondence is smooth. In fact, the standard way of introducing individual mistakes in the analysis of QRE is precisely with the help of a logistic function, just like the one used above. Thanks to this, the agents' payoff function no longer displays the step-wise character they have in the pure-majority game. The sources of such continuity are different. In the QRE model, it is the result of introducing a probabilistic error term in the choice of each agent, an error term due to individual traits and not connected to institutional considerations. In the model above, the continuity results from objective traits of the collective decision-making process.

One of the most interesting properties of QRE is that, unlike pure-strategy Nash equilibria, they offer comparative statics with respect to the payoff parameters. But, as I will show in the next chapter, all is not lost for Nash equilibria in this regard. By studying their stability sets we can obtain useful comparative statics results. In what follows I will rely on the method of stability sets as applied to Nash equilibria, not because I believe it has some inherent superiority with respect to QRE but because it is better suited for the goals of this book. The concept of QRE brings to the fore the role of uncertainty at the individual level whereas the analysis of stability sets I will develop later, instead, focuses on the role of changes in the structural payoffs. From a mathematical point of view, the method of stability sets can be considered a generalization of QRE. As Turocy (2005) makes clear, QRE describes the equilibrium of a game as it changes along a homotopy that starts in the centroid of the strategy space. Instead, as we will see in the next chapter, the method of stability sets considers all the possible homotopies, taking into account all the possible starting points (Medina 2013).

2.3.2 *Voters with Negative Costs*

As we already had the opportunity to discuss, the idea that, at least for some voters, the act of voting may not be costly at all, but may instead be a source of rewards predates the game-theoretic analysis of voting. In the context of a decision-theoretic model, this idea is problematic because the resulting theory cannot explain the connection between turnout and objective, structural parameters. But since there is no denying that such rewards exist, it is important to study how they can be integrated in a game-theoretic model.

As it happens, there is no difficulty in doing this. The model as it stands right now already allows for heterogeneity of costs and, therefore, nothing prevents us from postulating a distribution of said cost terms where some voters face a term $c < 0$. The main results do not depend on this.

One interesting anomaly of the decades-long debate around the turnout paradox is that, while the rational-choice model of voting is widely (and, as argued above, wrongly) perceived as a failure in accounting for the high *levels* of turnout we observe, it has succeeded in capturing many regularities about the *changes* in such levels. For instance, most studies confirm that closeness of an election tends to increase turnout, contrary to what one would expect in a model where voters use a different decision rule, such as the minmax strategy (Ferejohn and Fiorina 1974). This suggests that including voters with a negative cost can lead to a better model. After all, such voters could be the ones that largely determine the level of turnout and the margin of victory while those voters with a positive cost are the ones that cause the fluctuations of both variables.

This is a good modeling choice and, in fact, I will adopt it shortly. My reasons for doing such have mostly to do with analytical convenience and generality, though. We do not need psychic rewards to explain high levels of turnout. But, as we will see later, the introduction of these blocs of voters with negative costs allows us to cover in an analytically simple manner interesting cases of multiple equilibria, a topic of great interest in its own right.

2.3.3 *Differentiated Costs*

While the case for psychic benefits is convincing, it also suggests that the standard way of introducing them might be problematic. In the standard argument about "civic duty," voters experience some benefit from voting because they feel that by doing this they are fulfilling some obligation. But it is hard to see why these benefits would be independent of the outcome. Presumably, if citizens enjoy voting, they enjoy more winning. Whatever psychic or expressive rewards they obtain from voting, they must be larger if their preferred candidate wins. In fact, Anderson and Tverdova (2001) have recently corroborated this very point through studies of voters attitudes in several democracies.

So, one interesting modification would be to allow the cost of voting to vary with the outcome. As it happens, this kind of model retains the high-turnout equilibria that we have already obtained while also generating large, realistic margins of victory. To appreciate this point, let's modify the payoff functions of the model by introducing cost terms $c_{Si} < c_{Fi}$ so that now the expected value of each strategy is:

$$\begin{aligned} w_i(V_i, s_{-i}) = {} & (u_L(x_L) - c_{Si})g(V_L^i + 1, V_R) + \\ & (u_L(x_R) - c_{Fi})(1 - g(V_L^i + 1, V_R)), \end{aligned} \tag{2.7}$$

$$w_i(A_i, s_{-i}) = u_L(x_L)g(V_L^i, V_R) + u_L(x_R)(1 - g(V_L^i, V_R)). \tag{2.8}$$

This results in the following best-response correspondence:

$$\sigma_i^*(s_{-i}) = \begin{cases} 1 \quad \text{if } g(V_L^i+1, V_R|s_{-i})(\frac{\delta-(c_{Si}-c_{Fi})}{\delta}) > g(V_L^i, V_R|s_{-i}) \\ \qquad + \frac{c_{Fi}}{\delta}, \\ [0,1] \text{ if } g(V_L^i+1, V_R|s_{-i})(\frac{\delta-(c_{Si}-c_{Fi})}{\delta}) = g(V_L^i, V_R|s_{-i}) \\ \qquad + \frac{c_{Fi}}{\delta}, \\ 0 \quad \text{if } g(V_L^i+1, V_R|s_{-i})(\frac{\delta-(c_{Si}-c_{Fi})}{\delta}) < g(V_L^i, V_R|s_{-i}) \\ \qquad + \frac{c_{Fi}}{\delta}. \end{cases}$$

In this case, since the players' type is now a two-dimensional vector (c_{Si}, c_{Fi}), it would be slightly misleading to speak of a "cutoff point" equilibrium, no single type separates the group of citizens between voters and non-voters. But, as usual, we can characterize the equilibrium as a fixed-point of a function. To that end, let's define the following functions:

$$\gamma_L(V_L, V_R) = \#\{i : g(V_L^i+1, V_R)(1 - \frac{c_{Si}-c_{Fi}}{\delta}) > g(V_L^i, V_R) + \frac{c_{Fi}}{\delta}\},$$

$$\gamma_R(V_L, V_R) = \#\{j : g(V_L, V_R^j+1)(1 - \frac{c_{Sj}-c_{Fj}}{\delta}) > g(V_L, V_R^j) + \frac{c_{Fj}}{\delta}\}.$$

Now we can characterize an equilibrium for this new version of the game:

Theorem 3 *Let Γ be a voting game with outcome function defined by Eq. 2.1, with the payoff functions for player $i \in N_L$ defined by Eqs. 2.7 and 2.8 and analogous payoff functions for player $j \in N_R$. Then, the levels of turnout V_L^*, V_R^* constitute a Nash equilibrium of Γ if:*

$$\gamma_L(V_L^*, V_R^*) = V_L^*,$$

$$\gamma_R(V_L^*, V_R^*) = V_R^*.$$

Proof From the definition of γ_L and γ_R we can see that, at these levels of turnout, for all the players voting the expected payoff of voting is higher than that of abstaining. Likewise, those players not included in the sets measured by γ_L and γ_R are such that their expected payoff of abstaining is higher than that of voting. Thus, no player can benefit from a unilateral deviation and hence, the strategy profile is a Nash equilibrium.

In this version of the model, the expected payoff from abstaining remains the same as before; the change takes place in the expected payoff from voting. Intuitively, what this new function represents is a situation where, for the voter, the cost of voting decreases as the outcome of the election tilts more toward her preferred point. This seems plausible and realistic: there is no reason to believe that the cost of voting, as experienced by a citizen, is the same regardless of whether the outcome was a crushing defeat or a clamorous victory.

It is easy to see why this model can sustain equilibria with high margins of victory, at least for some parametrizations. First, the smoothness of the outcome function means that "surplus votes" are not really surplus: every vote pushes the outcome in the direction of the voter's preferences. As I already argued, this is consistent with the way electoral democracy works. Second, from the voter's point of view, voting mitigates its own cost, precisely by contributing to the scale of victory (or defeat). Thus, even if the mark of 50% plus one is met, a citizen still has reasons to vote in this model.

To appreciate how this added flexibility allows us to generate equilibria with high turnout and substantial margins of victory, let's look at another numerical example, with a similar setting to the preceding ones. Let $N_L = 6 \times 10^7$ and $N_R = 4 \times 10^7$. Given the high number of voters involved, a value $\rho = 4 \times 10^{-7}$ will adequately capture an election where there is at least a modicum of proportionality in the sense of affording the defeated party some role, no matter how modest, in shaping the outcome. Let's suppose that, for every citizen, the cost of voting contingent on success, the term c_S introduced above is indeed negative. That is, for citizens, voting for a victorious candidate is actually a reward. Even a small such benefit can support equilibria with large margins of victory. Assume, for instance that, for every voter $c_{Si}/\delta = c_{Sj}/\delta = -8.435 \times 10^{-8}$. As regards the cost of voting for a defeated candidate, let's assume a distribution such that, for all $j \in N_R$, $c_{Fj}/\delta < 10^{-5}$ and that, for 83.333% of citizens $i \in N_L$, $c_{Fi}/\delta < 10^{-5}$, then we can support a cutoff point equilibrium such that $(c_{FL}/\delta)^* = (c_{FR}/\delta)^* = 10^{-5}$ and where all citizens in N_R vote and 5×10^7 citizens in N_L vote: a turnout of 90 million voters and a margin of victory of 11.1%.

Of course this is just one example but I believe it carries several important lessons. First, as we have seen repeatedly, large electorates are compatible with high turnout for plausible values of the cost-benefit ratio. A value of $c_F/\delta = 10^{-5}$ may seem very small at first glance until we realize that it means that, if the cost of voting is \$1, the benefit of single-handedly deciding the election is \$100,000. Even with these values, this example shows that it is possible to obtain turnout levels similar to those of the world's largest democracies. Second, even relatively small expressive benefits for the supporters of the victor can explain large margins of victory. In this example, the cost of voting c_F is almost 60 times larger than the "expressive benefit" c_S. Third, and probably more important, although this model incorporates some degree of psychic benefits in the payoff function, they do not overwhelm the role of the other variables to the extent of making comparative statics impossible. Here the cost-benefit ratio c_F plays a decisive role which means that any of the observable changes that affect it, be it, for instance, changes in the registration laws, or changes in the ideological distance between the parties, will affect turnout.

2.4 Characterizing the Equilibria of Voting Games

The time has come to put together the ideas of the preceding analysis into a model that we can use to develop comparative statics results on aggregate turnout. To that end I will consider a highly general model of voting and will later introduce constraints that will make it easy to analyze.

The most important sense in which the model in this section generalizes the ones above is in its equilibrium concept. While up to this point I have described the Nash equilibria of voting games, now I will extend the analysis to cover correlated equilibria as well. The reason for this has nothing to do with the description of the equilibria themselves. None of the results in this section will overturn what we have already learned because every Nash equilibrium is also a correlated equilibrium. But for reasons that will become clear soon, especially in the next chapter, we will need to take our analysis beyond the equilibria themselves and consider the properties of their stability sets. When it comes to that, in games with arbitrarily large numbers of players the correlated equilibria generate results that are more useful than the ones that can be obtained by focusing only on the Nash equilibria. In due course I will elaborate this point.

Momentarily let's return to the case considered in the previous chapter (Sect. 1.2). A correlated equilibrium is a probabilistic measure μ defined over the game's strategy space such that no player benefits from a unilateral deviation. So, μ is a correlated equilibrium if:

$$\sum_{s_{-k}\in S_{-k}} \hat{u}_k(g(s), s_k)\mu(s) \geq \sum_{s_{-k}\in S_{-k}} \hat{u}_k(g(s_{-k}, s'_k), s'_k)\mu(s)$$

for every player k and every pair of strategies s_k, s'_k in k's strategy space. Since each player has two possible strategies, this condition describes a set of $2N$ inequalities, two for each player. To analyze them in more detail, let's introduce indicator variables $I(V_k|s)$ that will take value 1 if in strategy profile s player k votes and 0 otherwise. Thus, in a correlated equilibrium the following inequalities hold simultaneously for every player k:

$$\sum_{s\in S} \hat{u}_k(g(s), V_k)\mu(s)I(V_k|s) \geq \sum_{s\in S} \hat{u}_k(g(s_{-k}, A_k), A_k)\mu(s)I(V_k|s),$$

$$\sum_{s\in S} \hat{u}_k(g(s), A_k)\mu(s)(1 - I(V_k|s) \geq \sum_{s\in S} \hat{u}_k(g(s_{-k}, V_k), V_k)\mu(s)(1 - I(V_k|s).$$

In what follows I will return to the case of undifferentiated costs of voting so that the payoff functions $\hat{u}$ can be written as in the model with heterogenous voters so that for every i in N_L and every j in N_R:

- $\hat{u}_i(g(s), A_i) = u_i(g(s))$; $\hat{u}_j(g(s), A_j) = u_j(g(s))$,
- $\hat{u}_i(g(s), V_i) = u_i(g(s)) - c$; $\hat{u}_j(g(s), V_j) = u_j(g(s)) - c$.

With these values, and assuming again that both u_i and u_j are linear in x_L and x_R, the inequalities above become:

$$\sum_{s\in S}(u_i(x_L)g(V_L^i + 1, V_R) + u_i(x_R)(1 - g(V_L^i + 1, V_R)) - c)I(V_i|s)\mu(s) \geq$$

$$\textstyle\sum_{s\in S} u_i(x_L)g(V_L^i, V_R) + u_i(x_R)(1 - g(V_L^i, V_R))I(V_i|s)\mu(s),$$

$$\sum_{s\in S} u_i(x_L)g(V_L^i, V_R) + u_i(x_R)(1 - g(V_L^i, V_R))(1 - I(V_i|s))\mu(s) \geq$$
$$\textstyle\sum_{s\in S}(u_i(x_L)g(V_L^i + 1, V_R) + u_i(x_R)(1 - g(V_L^i + 1, V_R)) - c)$$
$$(1 - I(V_i|s))\mu(s),$$

for any i in N_L and

$$\sum_{s\in S}(u_j(x_L)g(V_L, V_R^j + 1) + u_j(x_R)(1 - g(V_L, V_R^j + 1)) - c)I(V_j|s)\mu(s) \geq$$
$$\textstyle\sum_{s\in S}(u_j(x_L)g(V_L, V_R^j) + u_j(x_R)(1 - g(V_L, V_R^j))I(V_j|s)\mu(s),$$
$$\sum_{s\in S} u_j(x_L)g(V_L, V_R^j) + u_j(x_R)(1 - g(V_L, V_R^j))(1 - I(V_j|s))\mu(s) \geq$$
$$\textstyle\sum_{s\in S}(u_j(x_L)g(V_L, V_R^j + 1) + u_j(x_R)(1 - g(V_L, V_R^j + 1)) - c)$$
$$(1 - I(V_j|s))\mu(s),$$

for any j in N_R.

Gathering terms and recalling that $\delta_i = u_i(x_L) - u_i(x_R)$ and $\delta_j = u_j(x_R) - u_j(x_L)$, these inequalities become:

$$\sum_{s\in S}(g(V_L^i + 1, V_R) - g(V_L^i, V_R))\mu(s)I(V_i|s) \geq \frac{c}{\delta_i}P(V_i|\mu),$$
$$\sum_{s\in S}(g(V_L^i, V_R) - g(V_L^i + 1, V_R))\mu(s)(1 - I(V_i|s)) \geq -\frac{c}{\delta_i}P(A_i|\mu),$$

for any i in N_L and

$$\sum_{s\in S}(g(V_L, V_R^j) - g(V_L, V_R^j + 1))\mu(s)I(V_j|s) \geq \frac{c}{\delta_j}P(V_j|\mu),$$
$$\sum_{s\in S}(g(V_L, V_R^j + 1) - g(V_L, V_R^j))\mu(s)(1 - I(V_i|s)) \geq -\frac{c}{\delta_j}P(A_j|\mu),$$

for any j in N_R.

Consider now the case of the two inequalities that must hold for an arbitrary agent i in N_1. There are only three possible ways in which they can be made consistent. In the first case, the player always votes so that $P(V_i|\mu) = 1$. Then, for every strategy profile s, $I(V_i|s) = 1$ so that the second equality trivially holds because both its sides take value 0. Denoting by E_μ the expected value of a variable under measure μ, we obtain then:

$$E_\mu(g(V_L^i + 1, V_R) - g(V_L^i, V_R)) \geq \frac{c}{\delta_i}.$$

Alternatively, the player never votes so that now it is the first inequality the one that holds trivially while the second one becomes:

$$E_\mu(g(V_L^i + 1, V_R) - g(V_L^i, V_R)) \leq \frac{c}{\delta_i}.$$

Finally, the measure μ may be such that in some profiles the player votes but not in others. In that case, both $P(V_i|\mu)$ and $P(A_i|\mu)$ take values strictly between 0 and 1. Then, the only way the two inequalities can be consistent is if:

$$E_\mu(g(V_L^i + 1, V_R) - g(V_L^i, V_R)) = \frac{c}{\delta_i}.$$

An implication of these inequalities is that, except for one particular value c/δ_i, all the voters choose pure strategies. This in turn implies that the distribution μ puts all the probabilistic mass on specific values V_L, V_R so that, from now on, we can treat any term $E_\mu(g(V_L, V_R))$ simply as $g(V_L, V_R)$.

These conditions describe the individual behavior of each agent given an equilibrium. Now we can aggregate these conditions to characterize total turnout. This we can do by noticing that each individual condition also helps us sort agents into voters or abstainers. Define

$$\gamma_L(V_L, V_R) = \#\{i : \delta_i \geq \frac{c}{g(V_L^i + 1, V_R) - g(V_L^i, V_R)}\},$$

$$\gamma_R(V_L, V_R) = \#\{j : \delta_j \geq \frac{c}{g(V_L, V_R^j) - g(V_L, V_R^j + 1)}\}.$$

So, an equilibrium is a pair of values V_L^*, V_R^* such that:

$$V_L^* = \gamma_L(V_L^*, V_R^*),$$
$$V_R^* = \gamma_R(V_L^*, V_R^*).$$

We have encountered similar conditions before. This is yet another consequence of the fact that every Nash equilibrium is also a correlated equilibrium.

In what follows I will focus on a special but useful case for which I will offer a more explicit characterization. As we have seen, an equilibrium depends on the aggregate distribution of payoff parameters. In keeping with the idea that for some citizens voting might not be a cost at all, and may instead be its own reward, I shall assume that, in addition to the players we have analyzed so far, each party also has some citizens for which $c \leq 0$. These groups will be labeled B_L and B_R. From the above definitions, we know that those players will always vote. Without any major loss of generality, let's assume also that for those citizens in N_L and N_R the parameters δ_i and δ_j are uniformly distributed with upper bounds δ^L and δ^R respectively. Since our interest here is mostly on the properties of the game's equilibria more than on the value of these equilibria, there is no harm in assuming further that $\delta^L = \delta^R = \bar{\delta}$, something that will greatly simplify the analysis. For notational convenience, we will denote $(g(V_L + 1, V_R) - g(V_L, V_R))$ as $D_L(V_L, V_R)$ and, analogously, $(g(V_L, V_R) - g(V_L, V_R + 1))$ as $D_R(V_L, V_R)$.

Under these assumptions, the value $\gamma_L(V_L, V_R)$ is:

$$\gamma_L(V_L, V_R) = \begin{cases} N_L & \text{if } \ c/D_L(V_L, V_R) < 0, \\ N_L - \frac{c}{\bar{\delta} D_L(V_L, V_R)} & \text{if } 0 < c/D_L(V_L, V_R) < \bar{\delta}, \\ 0 & \text{if } \ \bar{\delta} < c/D_L(V_L, V_R). \end{cases}$$

An analogous expression describes $\gamma_R(V_L, V_R)$. By definition, both $D_L(V_L, V_R)$ and $D_R(V_L, V_R)$ are greater than zero so we can eliminate the first case of these piece-wise functions. Since V_L and V_R enter symmetrically in the logistic outcome function, then it is also the case that $D_L(V_L, V_R) = D_R(V_L, V_R) = D(V_L, V_R)$.

A full description of the equilibrium requires the two equations we have established above. But we can simplify matters further if we simply focus on the margin of victory by combining the two equations into one. So, from now on I shall assume that:

$$m \equiv V_L - V_R,$$

$$b \equiv B_L - B_R,$$

$$n \equiv N_L - N_R.$$

As a result, we can describe the difference between the vote of the two parties as follows:

$$F(m) = \begin{cases} b & \text{if } \ \bar{\delta} < c/D(m), \\ b + \frac{n}{\bar{\delta}}(\bar{\delta} - c/D(m)) & \text{if } 0 < c/D(m) < \bar{\delta}. \end{cases}$$

Define $H(m, \rho) = (e^{-\rho m} + e^{\rho(m+1)} + e^{\rho} + 1)/(e^{\rho} - 1)$. Then, by applying the definition of g and putting the outcome function in terms of m, we obtain that:

$$F(m) = \begin{cases} b & \text{if } 1 < \frac{c}{\bar{\delta}} H(m, \rho), \\ b + n\left(1 - \frac{c}{\bar{\delta}} H(m, \rho)\right) & \text{if } \frac{c}{\bar{\delta}} H(m, \rho) < 1. \end{cases}$$

The fixed points of $F(m)$, that is the values of m such that $m = F(m)$ are the equilibria of the game. So we can now turn our attention to the properties of these equilibria, in particular, their comparative statics.

The function $F(m)$ is piece-wise and symmetric around 0 by which I mean that $F(m) = F(-m)$. It is easy to verify that there is a value $\tilde{m}$ such that if $|m| > \tilde{m}$, then $F(m) = b$. For values such that $|m| < \tilde{m}$, $F(m)$ is a concave function that reaches its maximum value at $m = 1/2$.

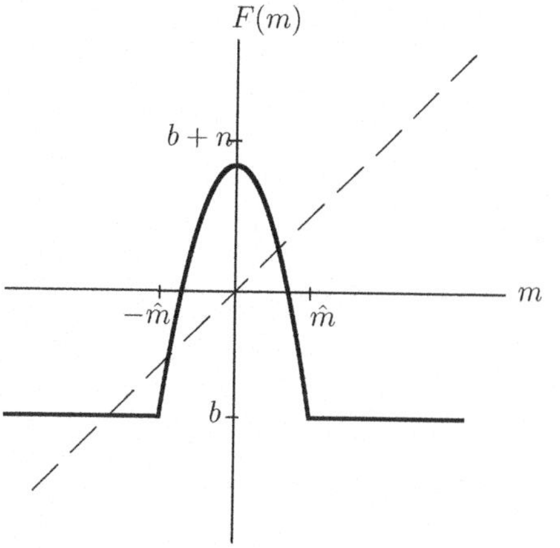

Fig. 2.1 A voting game with multiple equilibria: the intersections with the 45° constitute equilibrium levels of m

Ultimately, voting games are a "two-team" version of coordination games where each team has to overcome a threshold that, unlike what happens in "one-team" games, is endogenously determined by the other team. Therefore, it is not surprising that voting games inherit some properties of their simpler counterparts. One such property that will concern us here is the multiplicity of solutions.

Consider, for example, Fig. 2.1 that describes the behavior of $F(m)$ for some parameter values such that $b < 0$, $n > 0$, $b + n > 0$. As we can see, this game produces three fixed points for m: the game has multiple equilibria.

A numerical example may be of help here. If we take values $c = 1$, $\bar{\delta} = 10^6$ and $\rho = 5 \times 10^{-6}$, $b = -5 \times 10^5$ and $n = 2 \times 10^6$, there are three values of m that constitute equilibrium margins of victory: $m \approx 2.5 \times 10^5$, $m \approx -3.2 \times 10^5$ and $m = -5 \times 10^5$.

Since one of the central goals in this book is to obtain comparative statics results that connect turnout to the structural aspects of elections, this seems a major roadblock. When a game has multiple equilibria, different types of aggregate behavior are consistent with strategic rationality and, hence, it is not clear which such behavior to single out for analysis. The multiplicity of equilibria is often regarded as a source of embarrassment for the entire rational-choice enterprise since it is taken to mean that the underlying theory fails to give an accurate prediction.

I think such pessimism is unwarranted. It is true that if the equilibria themselves were the only information we could obtain from a game-theoretic analysis, then in the case of games with multiple equilibria all we would have at the end would be a list of possible solutions without any further element to decide what to do about them. But there is more we can learn about such solutions. In particular, it is possible to determine their stability sets as well, that is, the conditions under which we can expect each solution to emerge. This information gives us the missing pieces of the analysis so that we can infer comparative statics results even in games with multiple equilibria. In the next chapter I will turn to a thorough discussion of the concept of stability sets and how they can be applied to voting games.

Chapter 3
The Stability Analysis of Voting Games

3.1 Introduction

In the preceding chapter we established that voting games resemble the well-known "tipping games" of the literature on collective action. This resemblance is hardly surprising. After all, it is possible to conceive of voting games as a more general case of tipping games, one where the threshold level is endogenously chosen by a group of players with opposing interests. This suggests that as we proceed in our analysis themes familiar from the theory of collective action will emerge. After characterizing the solutions of voting games we encountered one such theme: multiplicity of equilibria.

We had already seen in Chap. 1 that the canonical, pure-majority voting games have a bewildering amount of equilibria. In that case those equilibria were, to a large extent, an artifact of the particular outcome function of those games. But the analysis of Chap. 2 shows that even with a more plausible outcome function voting games admit of many solutions.

Received wisdom claims that in these cases the theory must remain agnostic about the final result and confess to a lack of predictive power. But that is not the only road available. When applied to games with multiple equilibria, the method of stability sets calculates a probability distribution over the different equilibria. The most salient property of the resulting distribution is that it is a function of the game's payoffs. As a result, even in these games it is possible to make probabilistic predictions and to obtain comparative statics results that connect the changes in the predicted behavior with changes in the game's structural parameters.

In this chapter I will proceed in the following steps. First, I will rehearse briefly the foundations of the method of stability sets, explaining how it generates comparative statics results. To that end I will rely on the analysis of simple games with small numbers of players (Sect. 3.2). To implement the method of stability sets we need to establish the stability conditions of any equilibrium, something that I will do here with the help of the tracing procedure. Once the mathematical apparatus is

© Springer International Publishing AG 2018

L. F. Medina Sierra, *Beyond the Turnout Paradox*, SpringerBriefs in Political Science,
https://doi.org/10.1007/978-3-319-73948-9_3

in place, I will apply it to the study of the general voting games introduced in the preceding chapter (Sect. 3.3). This will be the central result that will enable the structural analysis of turnout I will present later. Here I will illustrate the use of the method with an example of how to obtain comparative statics with respect to changes in the voting costs (Sect. 3.4).

3.2 The Method of Stability Sets in a 2 × 2 Game

In much the same way as conventions and social mores, and for pretty much the same reasons, Nash equilibria are their own justification. To those of us raised under the metric system the English system of weights and measures feels rather quaint and inconvenient. But this has not prevented it from serving well the citizens of the most advanced industrial economy, that conduct their everyday trades using it. Whatever its merits and disadvantages, the English system remains widely used in the US, this in spite of conscious efforts to eliminate it. Doubtless, custom is the main reason. Once the people around you rely on it, there is little reason to go against the grain.

This property of equilibria explains why rational agents choose the strategies prescribed by an equilibrium profile provided the rest are doing so. But in the case of multiple equilibria this is not enough. After all, every equilibrium satisfies this same condition and they all result in different patterns of aggregate behavior. Strategic rationality is no longer enough to explain why a specific pattern emerges. For that we need to resort to the set of non-strategic beliefs and understandings shared by the players that serves to buttress their convergence around a specific equilibrium.

This is already a well-established reasoning in game theory and the method of stability sets transforms it into a source of probabilistic predictions. Intuitively, the method of stability sets endeavors to determine the scope of possible belief conditions that support a specific equilibrium. The larger such scope, that is, the more the belief conditions that will lead players in the direction of said equilibrium, the more we should expect that equilibrium to emerge from any given instance of the underlying game (Medina 2005).

Let's illustrate this analysis with an example of a 2×2 game. Consider the payoff matrix of Fig. 3.1.

We could take this game to represent a situation in which two agents can benefit if both of them cooperate in a specific plan. But, if one of them defects, the benefits will not materialize and, instead, the person that did cooperate will face a cost. This is a typical coordination game and I choose it precisely because it is already very well-known and widespread in its use.

Fig. 3.1 Payoff matrix of a coordination game

		2: C	2: D
1	C	$1, 1$	$-1, 0$
1	D	$0, -1$	$0, 0$

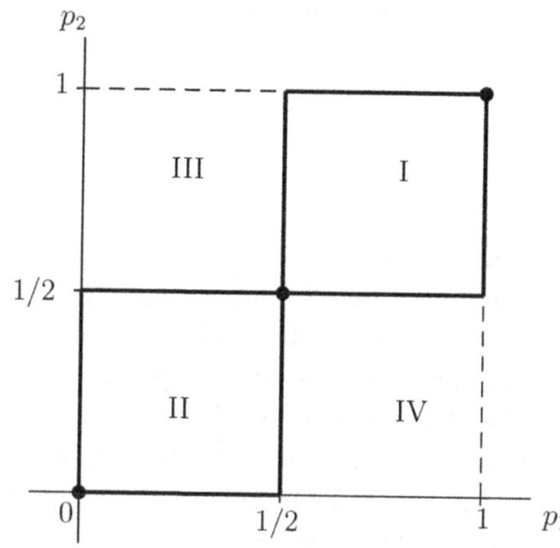

Fig. 3.2 Best-response correspondences of the coordination game

To calculate the Nash equilibria of this game, let's denote the probability that each player chooses C with p_1 and p_2 respectively. The expected payoff of the game for Player 1 is:

$$\begin{aligned} v_1(p_1, p_2) &= 1 \cdot p_1 p_2 + (-1) \cdot p_1(1 - p_2), \\ &= p_1(2p_2 - 1). \end{aligned}$$

From this payoff function we infer that the value of p_1 that maximizes Player 1's payoff is given by the following best-response correspondence:

$$p_1^*(p_2) = \begin{cases} 1 & \text{if } p_2 > 1/2, \\ [0, 1] & \text{if } p_2 = 1/2, \\ 0 & \text{if } p < 1/2. \end{cases}$$

An analogous expression represents Player 2's best-response correspondence. Putting both of them together, we obtain the following graph that shows the game's Nash equilibria (Fig. 3.2).

From this graph we deduce that the game has three Nash equilibria: (C, C), (D, D) and a mixed-strategy equilibrium where $p_1 = p_2 = 1/2$.

For the sake of argument, let's consider what would happen if each player had prior beliefs about the other player's behavior at, say, $p_1 = p_2 = 0.75$ and that these beliefs became common knowledge. Although neither one is absolutely sure about the other, they both assign a probability of 3/4 to their counterpart cooperating. The best-response correspondences tell us that, given those beliefs, which from now on I will call *initial belief conditions*, once they are common knowledge, both players will find it optimal to cooperate. In a sense still to be formalized, we can say that these belief conditions support the equilibrium C, C.

Fig. 3.3 A new coordination game

		2: C	2: D
1	C	2, 2	−1, 0
	D	0, −1	0, 0

The same is true for all the profiles of initial belief conditions in Region I. Hence, we say, following the terminology of Harsanyi and Selten (1988) that Region I is the *stability set* of the equilibrium C, C. Likewise, if a profile in Region II were to become common knowledge, the optimal strategies for both players would be to play the equilibrium D, D. Thus, Region II is the stability set of this other equilibrium. It is noteworthy that the mixed-strategy equilibrium is its own stability set in the sense that no other strategy profile will lead players to it.

This is a property of all mixed-strategy Nash equilibria. The intuitive reason for this is that, since they require players to be indifferent between their strategies, any tiny change in the beliefs about other players will tilt them in the direction of a pure strategy, that is, away from the mixed-strategy equilibrium.

In this example, Regions I and II are of equal size. This suggests that, in some sense, both equilibria are equally likely. This is not yet a rigorous statement in part because we have not made any claim or assumption about the probability with which any profile in any of these Regions emerges. If, however, for the sake of argument we suppose that all of them are equally likely, then we would be justified to claim that each of these two equilibria may emerge with a probability of 1/2. With different assumptions about the underlying distribution of initial belief conditions we would obtain different probabilities.[1]

For the purposes of this analysis, the exact magnitudes of these probabilities are not of interest so that I shall maintain the assumption of equal probability of initial beliefs. What matters is the way in which they respond to changes in the structural parameters, their comparative statics. In this regard, the stability sets have a very useful property: their size, and hence the probability of each equilibrium, is a function of the payoffs.

To see this, let's modify the previous example by increasing the benefit that both players obtain from the coordination equilibrium C, C. Now the payoff matrix is (Fig. 3.3).

Now the expected value of the game for Player 1 is:

$$\begin{aligned} v_1(p_1, p_2) &= 2 \cdot p_1 p_2 + (-1) \cdot p_1(1 - p_2), \\ &= p_1(3p_2 - 1). \end{aligned}$$

[1]In this paragraph I have ignored Regions III and IV but this is purely for expository reasons. Later, in the fully formal analysis of stability sets we will consider the entire strategy space.

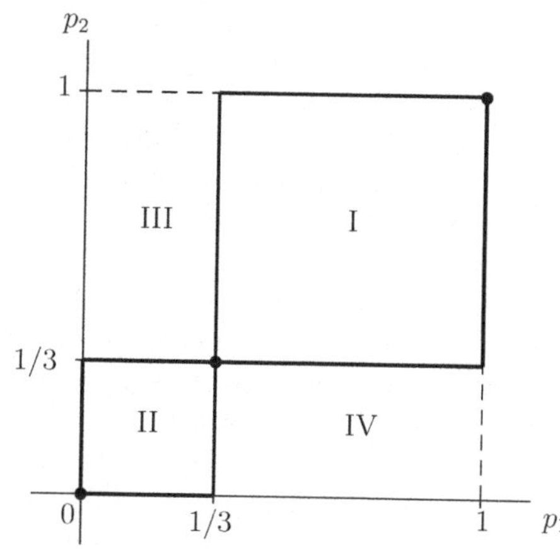

Fig. 3.4 Best-response correspondences of the new coordination game

As a result of this change in the payoffs, the best-response correspondences also change:

$$p_1^*(p_2) = \begin{cases} 1 & \text{if } p_2 > 1/3, \\ [0,1] & \text{if } p_2 = 1/3, \\ 0 & \text{if } p < 1/3. \end{cases}$$

Where, as usual, the best-response correspondence of Player 2 is analogous. Figure 3.4 shows us the resulting graph.

As we can see now, due to the increase in the payoff of the coordination equilibrium, the size of the stability set of C, C (Region I) has increased while Region II has shrunk. This is due to the fact that the unstable mixed-strategy equilibrium, which acts as the border between both regions, is a continuous function of the payoffs, unlike the pure-strategy equilibria. Whereas in the previous example both Regions were of equal area, now Region I is four times larger than Region II. In other words, with these new payoffs, coordination becomes more likely. Maintaining the assumption of equal probability among all the initial belief conditions, we can say that now the probability of C, C is 80%. In essence, this is the way in which the method of stability sets generates comparative statics.

So, the key is to compute the size of the stability set of each equilibrium in the game. In this 2 × 2 example this was simply a matter of plotting the best-response correspondences. But for more complicated games this procedure would be impractical. Furthermore, if we want to go beyond comparing particular examples of a given game, we need to be able to characterize the stability sets of the different equilibria for any general payoff vector.

There is a way to simplify this task. Suppose we can have a formula that tells us, for every given initial belief conditions, the stability set to which they belong, then we can apply this formula to the entire strategy space and obtain a rule to partition it among the different stability sets. The definition of a stability set already suggests how such a formula might look: for any arbitrary strategy profile, calculate the optimal reaction of every player. In simple games this works instantly as we can verify in the examples above. But with large numbers of players this method may not work because, as some players adjust their strategies when the initial belief conditions become common knowledge, they may lead the remaining players to a response different from the equilibrium they would have chosen under the initial profile.

This difficulty can be overcome by using an idea first developed by Harsanyi and Selten (1988): the tracing procedure. Instead of treating as instantaneous the adjustments the players' strategies would suffer as they react to the initial belief conditions, the tracing procedure phases in common knowledge so that said ajustments occur gradually. This means that, for every profile of initial belief conditions we obtain a path of equilibria as they become common knowledge. The end point of this path is, then, the equilibrium to which each initial belief condition is assigned.

Although the tracing procedure will not teach us much we do not already know about the 2×2 game above, it will be useful to illustrate how it works in this case. Let's return to the fist payoff matrix (Fig. 3.1). For reasons that will soon become clear, let's denote this original game as Γ^1. The first step of the tracing procedure is to create a replica of the original game where the players' payoffs depend on their own strategy and their beliefs about the remaining players. In this case, denote such initial belief conditions by q_1, q_2. Then, just as we obtained the game's expected payoff for each player, we can obtain it for the replica game which we shall now denote Γ^0:

$$
\begin{aligned}
v_1^0(p_1, q_2) &= p_1(2q_2 - 1), \\
v_2^0(q_1, p_2) &= p_2(2q_1 - 1).
\end{aligned}
$$

Intuitively, we could think of these payoff functions as describing a situation in which each agent is "playing" not against a strategic actor but against the prior beliefs she holds about the other actor. For instance, here the payoff of Player 1 depends on p_1 and the beliefs on Player 2, represented by q_2.

Now, let's introduce a parameter $0 < \lambda < 1$ that will identify the different members of a family of games Γ^λ. For each game, its payoff functions are:

$$
\begin{aligned}
v_1^\lambda(p_1, p_2, q_1, q_2) &= \lambda v_1^1(p_1, p_2) + (1 - \lambda) v_1^0(p_1, q_2), \\
&= \lambda p_1(2p_2 - 1) + (1 - \lambda) p_1(2q_2 - 1), \\
v_2^\lambda(p_1, p_2, q_1, q_2) &= \lambda v_2^1(p_1, p_2) + (1 - \lambda) v_2^0(p_1, q_2), \\
&= \lambda p_2(2p_1 - 1) + (1 - \lambda) p_2(2q_1 - 1).
\end{aligned}
$$

With this setup in place, we can now choose an arbitrary profile of initial belief conditions q_1, q_2 and then determine the equilibria for each game Γ^λ; this will generate the path of equilibria that will determine the stability set of this profile.

The game Γ^0 has a unique equilibrium because it is simply two juxtaposed decision problems, one for each player, each with a unique solution. In contrast, we know that Γ^1 has multiple equilibria. So, there will be only one continuous path of equilibria from Γ^0 to Γ^1 although some additional, incomplete paths appear along the way.

To illustrate the procedure, let's consider the case of initial belief conditions $q_1 = q_2 = 1/4$. Then, for any value λ, the profile $p_1 = p_2 = 0$ constitutes an equilibrium. We know that when $\lambda = 1$ the game has other equilibria but we can now be certain that the only continuous path is the one of (D, D). If, instead, we chose initial belief conditions $q_1 = q_2 = 3/4$, then the only equilibrium that exists for every value λ, that is, the only continuous equilibrium path is $p_1 = p_2 = 0$. So, the tracing procedure maps the first pair of initial belief conditions onto (D, D) and the second one onto (C, C).

For illustration purposes, I have left some loose ends. Once we analyze the specific case of large voting games I will present a fully formalized version of the analysis. In the meantime, we should notice that, just as we would have expected, the outcome of the path of equilibria along the parametrized family of games Γ^λ, and with it the outcome of the tracing procedure, depends upon the specific profile we choose for the initial belief conditions. So, thanks to the tracing procedure we can map each individual profile of initial belief conditions into a specific pure-strategy equilibrium.

3.3 The Stability Sets of Large Voting Games

The time has come to apply these ideas to the case of voting games. Once we apply the method of stability sets to these, we will have a solid foundation on which to build the analysis of the comparative statics of turnout, the ultimate goal of this project.

To that end, we will need to modify the definition of a voting game making it more general to accommodate the tracing procedure. We shall start by defining, for each player k a replica of the original strategy space denoted by Q_k. Just as we did for the strategy spaces S_k, we shall denote the set of lotteries in this new space by $\Delta(Q_k)$ and any arbitrary $\omega_k \in \Delta(Q_k)$ will represent the probability of V_k in this space. The function $\nu : Q \rightarrow [0, 1]$, a probability distribution over this replica, will represent the initial belief conditions. Since it is a distribution, it is subject to the usual constraints $\nu(q) \geq 0, \forall q$ and $\sum_{q \in Q} \nu(q) = 1$. Next, we introduce a parameter $0 \leq \lambda \leq 1$ to characterize a family of games Γ^λ. Each of these games is like the games we have been analyzing so far only that the payoff functions $w_k^\lambda : \times_{k=1}^N \Delta(Q_k) \times_{k=1}^N \Delta(S_k) \rightarrow \Re$ are more general than before:

$$w_k^\lambda(\mu, \nu) = \sum_{s \in S} \sum_{q \in Q} (\lambda \hat{u}_k(g(s), s_k) + (1 - \lambda)\hat{u}_k(g(q_{-k}, s_k), s_k))\mu(s)\nu(q).$$

The tracing procedure requires the computation of equilibria along the entire path λ. So, we need to extend the definition of a correlated equilibrium to the new setup. We will say that μ_λ is a correlated equilibrium of Γ^λ if

$$\sum_{s\in S}\sum_{q\in Q}(\lambda\hat{u}_k(g(s),s_k)+(1-\lambda)\hat{u}_k(g(q_{-k},s_k),s_k))\mu(s)\nu(q)\geq$$
$$\sum_{s\in S}\sum_{q\in Q}(\lambda\hat{u}_k(g(s_{-k},s'_k),s'_k)+(1-\lambda)\hat{u}_k(g(q_{-k},s'_k),s'_k))\mu(s)\nu(q)$$

for every player k and for every pair of strategies s_k, s'_k. In the same way as we did when characterizing the game's equilibria, let's introduce indicator random variables $I(V_k|s)$ for each k that take value 1 if Player k votes in strategy profile s and 0 otherwise, so the inequalities can be expressed as:

$$\sum_{s\in S}\sum_{q\in Q}(\lambda\hat{u}_k(g(s),V_k)+(1-\lambda)\hat{u}_k(g(q_{-k},V_k),V_k))I(V_k|s)\mu(s)\nu(q)\geq$$

$$\sum_{s\in S}\sum_{q\in Q}(\lambda\hat{u}_k(g(s_{-k},A_k),A_k)+$$
$$(1-\lambda)\hat{u}_k(g(q_{-k},A_k),A_k))I(V_k|s)\mu(s)\nu(q),$$

$$\sum_{s\in S}\sum_{q\in Q}(\lambda\hat{u}_k(g(s),A_k)+(1-\lambda)\hat{u}_k(g(q_{-k},A_k),A_k))(1-I(V_k|s))\mu(s)\nu(q)\geq$$

$$\sum_{s\in S}\sum_{q\in Q}(\lambda\hat{u}_k(g(s_{-k},V_k),V_k)+$$
$$(1-\lambda)\hat{u}_k(g(q_{-k},V_k),V_k))(1-I(V_k|s))\mu(s)\nu(q).$$

Just as we have been using the notation V_M and V_M^k to represent the total number of votes for party M in a given strategy profile and the same number without counting k's choice, we need an analogous notation to represent the total number of votes for a party in the replica space Q. This we will do by introducing W_M and W_M^k. The utility functions of the games under study here are the same as the ones we had already analyzed. So, after using all the properties of $\hat{u}_k$ we can rewrite these inequalities as:

$$\lambda E_\mu(g(V_L^i+1,V_R)-g(V_L^i,V_R))+$$
$$(1-\lambda)E_\nu(g(W_L+1,W_R)-g(W_L,W_R))\geq\frac{c}{\delta_i}P(V_i),$$

$$\lambda E_\mu(g(V_L^i,V_R)-g(V_L^i+1,V_R))+$$
$$(1-\lambda)E_\nu(g(W_L+1,W_R)-g(W_L,W_R))\geq\frac{c}{\delta_i}P(A_i),$$

for any player i in N_L and with analogous inequalities for any j in N_R. If we take again the steps we took in the previous chapter, we conclude that these inequalities generate a cut-off point that partitions the set of players in N_L into voters and abstainers. (The same holds for players in N_R.) So, the number of voters from N_L is:

$$\gamma_L^\lambda(V_L, V_R) = \#\{i : \delta_i \geq \frac{c}{\lambda D(V_L, V_R) + (1-\lambda)D(W_L, W_R)}\},$$

$$\gamma_R^\lambda(V_L, V_R) = \#\{j : \delta_j \geq \frac{c}{\lambda D(V_L, V_R) + (1-\lambda)D(W_L, W_R)}\}.$$

Therefore, we arrive at a description of the equilibria similar to the one we obtained above. For any game Γ^λ, an equilibrium is defined by levels of turnout $V_L^{*\lambda}, V_R^{*\lambda}$ such that:

$$V_L^{*\lambda} = \gamma_L^\lambda(V_L^*, V_R^*),$$

$$V_R^{*\lambda} = \gamma_R^\lambda(V_L^*, V_R^*).$$

For simplicity, let's focus only on the margin of victory $m = V_L - V_R$. Putting these two conditions together, we conclude that an equilibrium value $m^{*\lambda}$ is a fixed-point of $\gamma_L^\lambda - \gamma_R^\lambda = F^\lambda$. So, in equilibrium:

$$m^{*\lambda} = F^\lambda(m^{*\lambda}).$$

When $\lambda = 1$ this game is exactly identical to the original one so that, depending on the parameter values, it may have multiple equilibria. Instead, when $\lambda = 0$, there is only one equilibrium: a citizen k votes if $\delta_k > c/D(W_L, W_R)$ and abstains otherwise. Now we need to map those equilibria induced by initial belief conditions W_L, W_R onto one of the equilibria of the original game. For that we need to describe the path that results as λ increases.

To that end, let's study what happens to the path in a neighborhood of an equilibrium. Suppose that m^* is an equilibrium such that there is an interval $[\underline{m}, \bar{m}]$ that contains $\hat{m}$ such that $F^1(\underline{m}) > \underline{m}$ and $F^1(\bar{m}) < \bar{m}$. Since the payoff functions of Γ^0 do not depend on the strategies S but on initial belief conditions Q, $F^0(m)$ does not depend on m either. Assume that $F^0(m) = l$, a constant such that, for every m in the interval under consideration $l > F^1(m)$, and hence $l > m^*$. As λ goes from 0 to 1, $F^\lambda(m)$ goes continuously from $F^0(m)$ to $F^1(m)$. Then, every value of $F^0(m)$ is mapped onto a smaller value, including the fixed point: the equilibrium of Γ^0 is mapped onto m^*, the equilibrium of Γ^1. Likewise, if $l < F^1(m)$, then the equilibrium of Γ^0 will be mapped onto a higher value. Putting these two cases together, we conclude that the tracing procedure will map all the values of this interval onto m^*. In other words, all the values in this interval belong to the stability set of m^*. Figure 3.5 illustrates this situation.

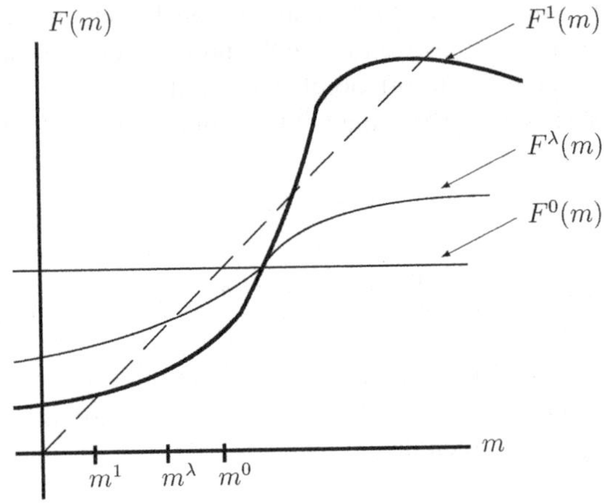

Fig. 3.5 The tracing procedure in action: as λ moves from 0 to 1, the F function approaches $F^1(m)$ and the equilibrium of the auxiliary game goes from m^0 to m^1

If instead we assume that $F^1(\underline{m}) < \underline{m}$ and $F^1(\bar{m}) > \bar{m}$, the situation is different. In this case, the values $m > \hat{m}$ are mapped onto even higher values and the values $m < m^*$ are mapped onto even lower values. The equilibria of Γ^0 on this interval drift away from m^*: the equilibrium m^* has no stability set.

This means that if we know the behavior of $F^1(m)$ in the neighborhood of an equilibrium we can easily know if said equilibrium is stable or not. We can even use this analysis to devise a simple criterion to pronounce on this matter. In the neighborhood of an unstable equilibrium $F^1(m) < m$ if $m < m^*$ and $F^1(m) > m$ if $m > m^*$. In turn, this means that, in that neighborhood, the slope of $F^1(m)$ is higher than the slope of the 45° line.[2] We can summarize this in an important result:

Theorem 4 *Let Γ_{GO} be a voting game with a generalized outcome function and let $m = V_L^* - V_R^*$ be the margin of victory for L generated by an equilibrium profile of this game. Then, m^* represents an unstable equilibrium if and only if:*

$$\left.\frac{dF^1(m)}{dm}\right|_{m=m^*} > 1.$$

Remark Conversely, if the equilibrium is stable the derivative of $F^1(m)$ will be less than one.

It would be tedious and unrewarding to try to describe all the possible patterns of equilibria in a voting game for arbitrary distributions of δ_k. It is better to turn our attention to the setup we have already described with the use of some mild assumptions. In this case, as happens so often with coordination games, there are three equilibria. Thanks to the preceding analysis we can now determine that two

[2] Results of this kind are familiar in stability analysis. An example similar in spirit can be found in Jackson and Yariv (2007).

of those equilibria, the lowest and the highest equilibrium values of m are stable, whereas the intermediate value is unstable, which is to say, its probability is zero.

We can exploit this pattern to obtain comparative statics over the probability distribution of the game's stable equilibria. Since the unstable equilibrium responds to changes in the structural parameters, we know that anything that moves it to the left will reduce the stability set, and hence the probability, of the low equilibrium and, conversely, anything that moves it to the right will increase that same probability.

3.4 An Example: Comparative Statics of Voting Costs

To fix ideas about how the method of stability sets can help us to generate comparative statics in voting games with multiple equilibria, let's consider for a moment a simple mental exercise: what would happen if, for exogenous, unspecified reasons, the cost of voting for the supporters of one party goes up? This is not a highly realistic example and, in fact, I chose it only for heuristic reasons. After all, usually the cost of voting for any citizen depends on reasons unrelated to her ideological preferences. That said, there are real-life instances that come close to this situation. In the American South before the Voting Rights Act the segregationist regime put in place several measures, such as poll taxes and manipulated literacy tests, calculated to make it harder for black citizens to vote. Given the realities of the moment, it would have been disingenuous to claim that these restrictions were unrelated to the citizens' political preferences.

While in the next chapter I will develop a more nuanced model, for the time being let's simply consider a situation where the citizens face voting costs depending on their party allegiances and let's study what happens as those costs change. So, we can modify the voting game we have analyzed so far by introducing voting costs c_L and c_R. Now, let's see what is the result of a change in c_L.

We know that the equilibrium conditions of the game are:

$$V_L^* - \gamma_L(V_L^*, V_R^*) = 0,$$
$$V_R^* - \gamma_R(V_L^*, V_R^*) = 0,$$

where the functions γ_L, γ_R are defined as:

$$\gamma_L(V_L, V_R) = \#\{i : \delta_i \geq \frac{c_L}{g(V_L^i + 1, V_R) - g(V_L^i, V_R)}\},$$
$$\gamma_R(V_L, V_R) = \#\{j : \delta_j \geq \frac{c_R}{g(V_L, V_R^j) - g(V_L, V_R^j + 1)}\}.$$

If we want to find the effect of a change in an exogenous parameter, we can apply the Implicit Function Theorem to the equilibrium conditions. So, taking the differential of the equation for V_L we see that:

$$\left(1 - \frac{\partial \gamma_L}{\partial V_L}\right) dV_L - \frac{\partial \gamma_L}{\partial c_L} dc_L = 0,$$

$$\frac{dV_L}{dc_L} = \frac{\frac{\partial \gamma_L}{\partial c_L}}{1 - \frac{\partial \gamma_L}{\partial V_L}}.$$

There is an analogous expression for V_R but since γ_R does not depend on c_L, the numerator of the equivalent expression becomes zero.

By the definition of γ_L we know that $\partial \gamma_L / \partial c_L$ is always negative. Therefore, the sign of the derivative dV_L/dc_L is positive if the denominator is negative and vice versa. This is the same as saying that dV_L/dC_L is positive if $\partial \gamma_L / \partial V_L > 1$.

This is the same condition that describes an unstable equilibrium. So, an increase in the cost of voting for L decreases the amount of votes for L in a stable equilibrium and increases them in an unstable equilibrium. The counterintuitive behavior of the unstable equilibrium is typical and, far from being a problem for our analysis, strengthens it. In fact, we know that said equilibrium is the border between the stability sets of the stable equilibria. As the cost of voting c_L increases, the margin of victory m of the unstable equilibrium increases, which means that the stable equilibrium with the high m is now less likely. In other words, the increase in c_L means that, in any observed election the party L will receive less votes and, furthermore, the likelihood of the outcomes favorable to L drops: raising c_L does an unmitigated damage to the party L.

None of this is surprising. In fact, it is fairly obvious. One suspects that segregationist politicians in the American South had already seen through this reasoning without the help of any game theory.

But, for all its commonsensical nature, this analysis has far-reaching theoretical consequences that become more clear when we take stock of the preceding chapters. In a nutshell, what at first glance seemed an absurd theory, incapable of explaining the most basic fact of turnout, the fact that large numbers of people vote, has morphed into a plausible source of realistic and falsifiable predictions.

Once we notice that the paradox of turnout is not an all-pervading problem for voting games, it becomes possible to formulate games that enable us to analyze the impact over electoral participation of changes in structural parameters. Just as coordination games, their simpler relative, voting games are prone to have multiple equilibria, something that may seem an impediment to any further progress. But thanks to the method of stability sets we can turn this crisis into an opportunity by calculating a probability distribution over all the game's equilibria that responds to the same structural changes we set out to analyze at the beginning. As the example just presented shows, this procedure allows us to generate comparative statics that are at once intuitive and rigorous.

Taken in isolation, the comparative statics result of this example does not take us very far. It uses a relatively elaborate apparatus to reach an obvious conclusion in an unrealistic case. But it is a start that suggests that the same apparatus, if applied to other cases may help us reach conclusions that would otherwise elude pure common sense. In the next chapter I will suggest one such possible use of the method by applying it to a substantive problem in the theory of electoral participation, the socioeconomic bias in turnout, and show how it can be the result of deep structural features of some modern democracies.

Chapter 4
Electoral Participation Bias and the Welfare State

For all the attention that levels of turnout receive in the game-theoretic literature, a case can be made that the most important problem in this regard is not the level of turnout but its bias. In principle, if every segment of the populace voted at the same rates, we could say that the lower the turnout the better; we could obtain the same degree of representation with a cheaper electoral exercise. But in reality this is not the case. There is massive evidence that, in country after country, electoral participation is biased so that those citizens at the top of the socioeconomic ladder vote at higher rates than the rest (Lijphart 1997).

Although this topic has not attracted enough attention in the rational-choice tradition, some contributions stand out. Feddersen and Pesendorfer (1997, 1999), in their work on the "swing-voter's curse" argued that the reason for the correlation between formal education and turnout is that, in the face of electoral uncertainty, poorly informed citizens strategically abstain from voting, lest their effect on the result ends up being deleterious to their own interests. In spite of the mathematical elegance of this approach, a true exemplar in its genre, the substantive argument has some difficulties because it assumes precisely what ought to be explained. In fact, it is not clear why schooling would be a predictor of voting in the first place, unless we assume that formal education is the sole, or preeminent source of political information, a point made long ago by Burnham (1982) with regards to the US. For several decades after World War II, in high-turnout Europe the correlation between schooling and voting was much weaker than in the US. This was most likely because the European parties at that time were much more active than their American counterparts at disseminating political information so that, even voters with limited exposure to formal education could think of themselves as informed.

Filer et al. (1993) have proposed another model, this one with a much more explicit politico-economic structure, where taxation is determined as a result of the political process, in line with the long-standing tradition of spatial theories of voting. To circumvent the "paradox of voting" she proposes a group-based solution concept

© Springer International Publishing AG 2018
L. F. Medina Sierra, *Beyond the Turnout Paradox*, SpringerBriefs in Political Science,
https://doi.org/10.1007/978-3-319-73948-9_4

where groups of voters are compensated by their respective political entrepreneurs. In a last analysis, Morton claims that the socioeconomic bias of turnout results from the poor facing higher costs of voting, in terms of opportunity costs of time at the polls and of information-gathering activities, than the rich. But this solution also begs some important questions. It is not clear that why the opportunity cost of voting for the poor will be higher than for the rich, given that the physical act of voting has such low costs that any tiny factor (e.g. traffic, weather, errands and family outings) can shift the value in one way or another. Likewise, to place the explanatory burden on the costs of information gathering incurs in the same mistake of Feddersen and Pesendorfer of reifying formal education as the only source of political knowledge against the evidence of many mass parties during the twentieth century.

Here I will propose another hypothesis. I argue that the socioeconomic bias in turnout results from structural features of contemporary economies, in particular, from their interaction with the politics of redistribution. Before presenting the model, I shall discuss the intuition behind it.

To that end, let's consider, for purely heuristic reasons, a simple and unrealistic case of political redistribution, at first glance unrelated to the problem at hand but that will help us sharpen our insight. Imagine a polity made of two regions A and B. The citizenry is supposed to vote on a tax proposal that will tax all citizens of A in an equal amount and transfer the proceeds to B where they will be evenly distributed among all those who live there. We would not be surprised if the proposal is opposed vehemently by the inhabitants of A and supported by those of B so that, at the end, everything comes down to which side has a larger population. This assumes that every citizen votes and we could complicate things a bit more if we relax this assumption. As we have already seen from our analysis in the preceding chapters, if voting is costly, there will be a threshold value of the cost-benefit ratio below which everyone will vote and above which no one will.

Now the outcome is not simply determined by a head count in each region because the final turnout will depend on the cost-benefit ratios in both of them. The region with more citizens below the threshold will generate more votes and win. As we have already seen, several attempts at understanding turnout in the rational-choice tradition emphasize the role of the voting cost but I do not regard this as a promising avenue of inquiry. The game-theoretic model of voting shows that turnout depends not just on costs but on the entire cost-benefit ratio so it is better to focus on the side of the benefits, which respond in systematic ways to properties of the economic structure.

In the case we are discussing, this suggests that the relative turnout in both regions will depend on how much the citizens stand to gain or lose from the proposal. For instance, if the residents of A are significantly richer than those of B and individuals have a diminishing marginal utility of income, it would be reasonable to conclude that more citizens in B will find it worth it to vote than those in A.

Let's suppose that for some unspecified reason, the transfer from A to B is stipulated to be made in kind. So, citizens of B will not receive money from A, but instead some specific good, say peanuts. In this case, more considerations will come into play in determining the policy outcome. For instance, citizens of B who are allergic to peanuts are bound to abstain so the vote's outcome will need to consider how big a segment of B's population they are.

Although a contorted metaphor with no claim to realism, this example captures one important aspect of our subsequent analysis. Food allergies are not usually a potent political force but the point remains that anything that affects the marginal utility that citizens obtain from a policy will affect their willingness to vote for it.

The standard models of electoral redistribution in the tradition of Meltzer and Richard (1981) analyze the electoral determinants of tax policy assuming that governments levy taxes and then distribute the proceeds among the citizens. For some purposes this is a reasonable modeling choice but it neglects an important aspect of the modern welfare state: although the government collects revenue from tax money, most distributive programs work by transferring goods, not money, to their beneficiaries. So, much like in our preposterous example of peanut transfers, in a welfare state net payers and net receivers face different budget constraints and different marginal utilities. For net payers, that is, agents with income above average the government's programs reduce their disposable income. For net receivers, instead, these same programs increase, not their income across the board, but their access to some specific goods, those provided by the government (e.g. housing, health, education). Typically, those goods cannot be transformed into income. Beneficiaries of the welfare state are not supposed to resale their handouts in the market. For the most part this is physically impossible (as in the case of hospital services) and on occasions even illegal. But, from the consumer's point of view, these goods and services are subject to marginal diminishing returns.

All else being equal, the stronger the effect of these marginal diminishing returns, the lower the benefits for recipients of redistribution from any marginal increase in the tax rate. This in turn will depress turnout among this segment of the electorate. Much like the citizens afflicted by peanut allergy in our fictitious example, the low-income voters in this model find themselves in a situation where additional goods-targeted redistribution does not contribute as much to their welfare as it would if it took the form of income. This asymmetry between beneficiaries and payers of redistribution has downstream effects in the electoral competition because it confronts parties who favor redistribution with a starker trade-off than the one their opponents face. Any increase in taxes mobilizes the beneficiaries but alienates the payers. So, if redistribution is targeted into goods with very low marginal utility for consumers, the tax increase a party would need to propose to mobilize low-income voters is too large to be worth the resulting electoral cost. Going back to our previous example, in the limit, when no citizen of B benefits from an increase transfer of peanuts, it stands to reason that no party will propose any kind of redistribution and will instead try to cultivate the constituency of A.

Whatever the merits of this insight, the usual models of voting are not equipped to scrutinize it because they assume universal voting. No doubt, this has been a reaction to the commanding presence of the turnout paradox. If we take the paradox to its last consequences, any model that does not rely on universal voting will have to settle for zero voting, which will render futile the entire analysis.

With the help of the models analyzed so far we do not need to face this dilemma. The game-theoretic model of voting can account not only for high levels of turnout, but also for its variation as the costs and, more importantly, the benefits vary. In this chapter I will illustrate this by developing a model of electoral competition where the structure of political redistribution affect the citizens' costs and benefits of voting and, hence, affect the process of policy-making.

4.1 The Model

4.1.1 The Economic Environment

Consider an economy where each agent i independently generates some personal income y_i. The state levies taxes at a flat rate τ for redistribution purposes. This is the basic set up originally studied by Meltzer and Richard (1981) and that is the subject of many other studies on the political economy of income distribution. But in this chapter's model I will introduce a seemingly small but significant variant. The standard model assumes that the tax levy is distributed among all the population as a cash transfer. This ignores the fact that most welfare state programs in fact distribute some specific goods, often only to eligible citizens. So, I will assume that the tax revenue is used by the government in the production of some good z. The process whereby a given amount of income y is transformed into a given level of output z is described by the production function $z = h(y)$. In keeping with standard models, this function will be increasing and concave ($h' > 0, h'' < 0$) so that there are diminishing marginal returns in the production of z. I will not assume that the only producer of z is the state. In fact, I will allow for the possibility that agents can "top off" their handout of z with their income. This is a realistic assumption: in a market economy, for most of the social services provided by the government (e.g. education, health, childcare, housing) there are substitutes or improvements privately produced that are available to agents with enough income to pay for them. In fact, this will be the key difference across agents in this model. The rest of the goods, which agents procure in the private markets, are aggregated in a consumption bundle x. The agents' preferences over these goods are represented by the Cobb-Douglas utility function $u(x, z) = \alpha \log x + (1 - \alpha) \log z$. In this analysis the relative price of goods will not play any role so we can simply normalize prices to 1.

To determine the budget constraint of a given agent we need to know how much is the handout of z provided by the government. Let's denote the average income of the economy as $\bar{y}$. Then, the total per capita tax revenue is $\tau \bar{y}$ so that, imposing a balanced-budget constraint, each agent receives a handout equivalent to $h(\tau \bar{y})$.

4.1.2 The Political Environment

Since we are interested in the political ramifications of this economic structure, let's assume that the tax policy τ is decided through political competition among parties. For simplicity, I will assume that there are two parties with ideological preference, that I will denote with the labels L and R and that compete for votes by proposing tax policies τ_L and τ_R: the policy space is one-dimensional.

Just as in much of the literature of spatial competition since Wittman (1983) and Roemer (1997), I will assume that the two parties of this model differ in their ideological orientation and can be thought of as representing a specific income level. I will assume that the preferences of party L correspond to those of a citizen with income y_L and, likewise, that party R represents an individual with income y_R. I will further assume that the median income is y^M and that $y_L < y^M < y_R$ so that the two parties straddle the median. For simplicity, I will assume that both parties are risk neutral.

Citizens are the other political actor we need to introduce in our model. For that purpose I will build on what we have already accomplished in the previous chapters, that is, I will assume that citizens evaluate the proposals from the two parties and decide which one to support based on their c/δ ratio, where c is simply the cost of the act of voting and δ is the benefit the citizen would receive if she were single-handedly to decide the election.

So far we have been treating voting costs as if they were identical for every voter. This is unrealistic and, curiously, complicates some mathematical expressions in the model. Instead, we obtain simpler, better-behaved terms if we assume that the cost of voting varies across individuals. We can model this by saying that citizen i's voting cost is a variable c_i. For simplicity, let's assume that c_i follows a uniform distribution $U[0, \bar{c}]$. In our analysis, the costs of voting do not play a substantial role in our results; all the action occurs on the side of the δ_i terms.

4.1.3 A Stage-by-Stage Analysis

The game proceeds in three stages: first the parties choose their policy platforms τ_L and τ_R, then the citizens vote based on those proposals and finally a policy is determined based on the vote tally. Following the standard procedure of backward induction, we must then calculate first the citizens' strategies, that is, we need to know whether they will vote or not, and for whom will they do if they so decide. Once we know this, we can use the result to determine the policy outcome of the election for any arbitrary pair of platforms. Finally, we need to know the policy proposals that the parties will propose so as to optimize their expected payoff. Since the model is quite general, we will not obtain specific expressions for the solution but will offer some general properties that characterize it.

Determining the Policy Outcome

In keeping with the previous chapters, I shall assume a continuous outcome function so that the final policy τ^* is determined as $\tau^* = \tau_L g(V_L, V_R) + \tau_R(1 - g(V_L, V_R))$ where $g(V_L, V_R)$ is the logistic function we have used throughout the previous chapters. With this outcome function, and given that there are only two candidates, we can restrict the analysis to "sincere" voting rules for each citizen, that is, if a citizen votes, she will do so for her preferred platform.

Citizens' Preferences

The citizens of the polity are at the same time economic agents. So, we infer their policy preferences from the way in which these policies benefit or hurt them in the economic sphere. To characterize each citizen's ideal policy, we first solve her utility maximization problem both with respect to the economic parameters she takes as a given and then, treating those parameters as a policy choice over which she has control: first find each consumer's optimal bundle as a function of income and taxes and then optimize the consumer's utility with respect to taxes.

Since the redistributive policy in this model works by transferring goods that agents cannot resell in the market, their budget constraint, unlike what happens in the textbook case, is non-linear. The government-provided handout of z makes sure that the consumer's optimum never has a consumption of z below $h(\tau\bar{y})$: such level is guaranteed without any further expense. As the graph in Fig. 4.1 shows, the consumers' budget constraint in this model has a "kink" so that, for some agents, the optimum is at the corner solution $(y_i(1-\tau), h(\tau\bar{y}))$. We need to calculate the income of such consumers because their ideal policy will be governed by an indirect utility function different from that of the rest.

An agent's after-tax income is $y_i(1-\tau) + h(\tau\bar{y})$. From the standard optimization of a Cobb-Douglas function we know that, unencumbered by the non-linearity of the budget constraint, a consumer's optimal bundle would be:

$$x_i^* = \alpha[y_i(1-\tau) + h(\tau\bar{y})],$$
$$z_i^* = (1-\alpha)[y_i(1-\tau) + h(\tau\bar{y})].$$

Instead, for voters whose optimal consumption is located at the corner solution, the bundle is:

$$x_i^* = y_i(1-\tau),$$
$$z_i^* = h(\tau\bar{y}).$$

So, the consumers at the corner solution are those for which the following two inequalities hold at the same time:

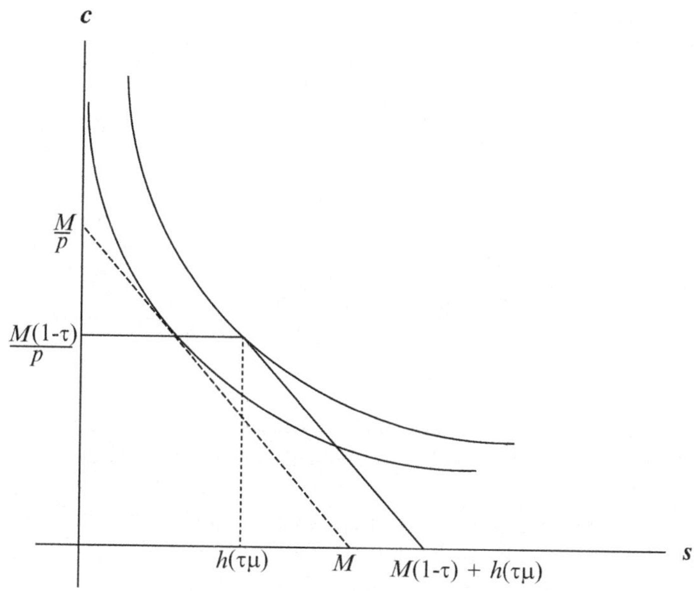

Fig. 4.1 Budget constraint for "Low-Income" consumers. The dotted line represents the pre-tax budget constraint

$$y_i(1-\tau) \leq \alpha[y_i(1-\tau)+h(\tau\bar{y})],$$

$$h(\tau\bar{y}) \geq (1-\alpha)[y_i(1-\tau)+h(\tau\bar{y})].$$

Both inequalities are valid for any citizen with an income level $y_i \leq \alpha/(1-\alpha)h(\tau\bar{y})/(1-\tau)$.

Denote by $V(y,\tau)$ the agent's indirect utility function, that is, the utility function evaluated at the optimal consumption bundle and represented in terms of the exogenous parameters. The preceding calculations imply that the indirect utility function is:

$$V(p,y,\tau) = \begin{cases} \log[y_i(1-\tau)+h(\tau\bar{y})]+K(\alpha) & \text{if} \quad y_i \geq \frac{\alpha h(\tau\bar{y})}{(1-\alpha)(1-\tau)}, \\ \alpha\log[y_i(1-\tau)]+(1-\alpha)\log h(\tau\bar{y}) \text{ if} & y_i \leq \frac{\alpha h(\tau\bar{y})}{(1-\alpha)(1-\tau)}, \end{cases}$$

where $K(\alpha) = \alpha\log\alpha + (1-\alpha)\log(1-\alpha)$, a constant that can be ignored for our purposes.

From this utility function we can compute the optimal tax rate for any citizen, depending on her income. But before doing so, we need to address some apparent

Fig. 4.2 Optimal tax policy as a function of income

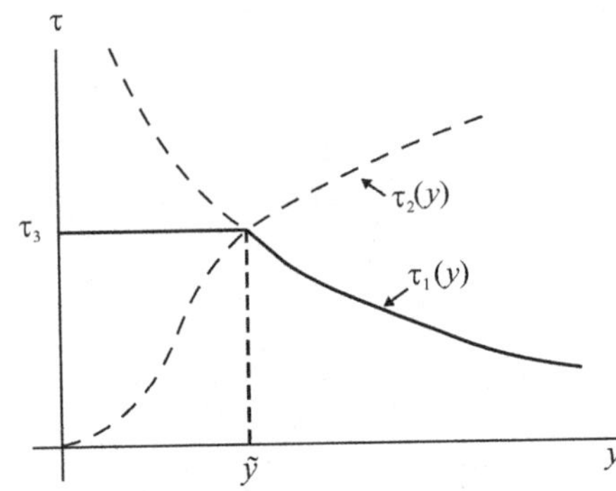

circularity in the problem. To establish the optimal tax rate we need to know which is the relevant budget constraint but the budget constraint, in turn, depends on the tax rate.

Define the implicit function $\tau_2(y)$ as the tax rate τ that solves the equation $y_i = \alpha h(\tau\bar{y})/(1-\alpha)(1-\tau)$. From its definition it is clear that τ_2 is increasing in y. Intuitively, $\tau_2(y)$ is the tax rate such that, if an agent has income y, her optimal consumption is at the corner of her budget constraint. So, we can obtain the optimal tax by maximizing the indirect utility function as follows:

$$\tau^*(y_i) = \begin{cases} \tau_1 : h'(\tau_1\bar{y}) = \frac{y_i}{\bar{y}} & \text{if } y_i \geq \frac{\alpha h(\tau_2(y_i)\bar{y})}{(1-\alpha)(1-\tau_2(y_i))}, \\ \tau_3 : \frac{h'(\tau_3\bar{y})(1-\tau_3)}{h(\tau_3\bar{y})} = \frac{\alpha}{\bar{y}(1-\alpha)} & \text{if } y_i \leq \frac{\alpha h(\tau_2(y_i)\bar{y})}{(1-\alpha)(1-\tau_2(y_i))}. \end{cases}$$

A major difference between these two solutions is that τ_3 is constant with respect to y whereas τ_1 is a decreasing function of y. This is because τ_3 is the optimal tax for consumers at the corner solution of their budget constraint. From their standpoint, the optimal tax is the one that brings them to the optimal consumption of z. Denote by $\tilde{y}$ the income level such that $\tau_1(\tilde{y}) = \tau_2(\tilde{y})$. From these two functions' definitions we know that the value $\tilde{y}$ is the income level where $h'(\tau\mu)\mu = (\alpha h(\tau\mu))/[(1-\alpha)(1-\tau)]$. But this is also the condition satisfied by τ_3. The following graph (Fig. 4.2) summarizes the situation and makes clear that this optimization problem partitions the electorate cleanly into two segments: one, with income less than $\tilde{y}$ whose optimal tax is fixed at τ_3 and another, with income above $\tilde{y}$ whose optimal tax is a decreasing function of income.

Intuitively, what this partition means is that the costs and benefits of this economy's distributive policy affect the citizens differently, depending on their income level. For citizens whose after-tax consumer optimum is at an interior solution, marginal changes in the tax level are reflected into changes in their disposable income. Instead, for citizens whose income places them at corner solution, marginal changes in the tax level affect their consumption of good z, which is subject to marginal diminishing returns.

Voting Choices

The citizens' preferences are the basis to calculate the voting decisions that go into determining V_L and V_R. In the previous chapters we have already developed the general framework so our task now is simply to adapt it to the model at hand. Up to now we have assumed that the benefit terms δ, so important in determining the voting strategies, are exogenously given. Now we can improve upon that. For any citizen i that prefers policy τ_L to τ_R, let $\delta_i(\tau_L, \tau_R) = V(y_i, \tau_L) - V(y_i, \tau_R)$. Conversely, for a citizen j with opposite preferences, $\delta_j(\tau_L, \tau_R) = V(y_j, \tau_R) - V(y_j, \tau_L)$. Define as before the functions γ_L and γ_R as:

$$\gamma_L(V_L, V_R) = \#\{i : \delta_i(\tau_L, \tau_R) \geq \frac{c_i}{g(V_L^i + 1, V_R) - g(V_L^i, V_R)}\},$$

$$\gamma_R(V_L, V_R) = \#\{j : \delta_j(\tau_L, \tau_R) \geq \frac{c_j}{g(V_L, V_R^j) - g(V_L, V_R^j + 1)}\}.$$

If we let $D(V_L, V_R) = g(V_L + 1, V_R) - g(V_L, V_R)$ and use the fact that the individual cost terms c_k follow a uniform distribution, then, for any arbitrary level of income y_k the proportion of citizens that vote is $D(V_L, V_R)\delta_k(\tau_L, \tau_R)$ multiplied by the normalizing constant $1/\bar{c}$.

Later we shall see that in any equilibrium $\tau_L > \tau_R$ so we can use that as an assumption. Then, there is a specific income level y^0 such that every citizen with income $y_i < y^0$ will prefer τ_L to τ_R and the opposite will be true for those with income $y_j > y^0$. This is a consequence of the fact that, for any citizen, the ideal tax policy is a decreasing function of income.

If we treat income y as a continuous variable, aggregating all the voters across income levels γ_L and γ_R becomes.

$$\gamma_L(V_L, V_R) = \frac{1}{\bar{c}} \int_0^{y^0} D(V_L, V_R)\delta_i(\tau_L, \tau_R)\, dy_i,$$

$$\gamma_R(V_L, V_R) = \frac{1}{\bar{c}} \int_{y^0}^{\infty} D(V_L, V_R)\delta_j(\tau_L, \tau_R)\, dy_j.$$

As usual, the equilibrium of the voting game is given by values V_L^*, V_R^* such that:

$$V_L^* = \gamma_L(V_L^*, V_R^*),$$
$$V_R^* = \gamma_R(V_L^*, V_R^*).$$

For simplicity, we can reduce these two equations to one and express the equilibrium in terms of the margin of victory (or defeat) of the L party by defining $m = V_L - V_R$. Then, if we define $F(m) = \gamma_L(m) - \gamma_R(m)$, we are back to the case already studied and the game's equilibrium is defined as the fixed point of F, that is, the solution to the equation

$$m^* = F(m^*).$$

One unfortunate fact of the analysis of large games is that without specific assumptions about the underlying distributions of players' payoffs, it is impossible to know how many equilibria the game will have. This case is no exception. But the results of the stability analysis of Chap. 3 allow us to obtain general conclusions of comparative statics even if the game has multiple equilibria.

There is no significant loss of generality if we assume that the function F has three fixed points that I will denote as $\underline{m}, m^u, \bar{m}$. Furthermore, I will assume that $\underline{m} < m^u < \bar{m}$ and that m^u is the unstable equilibrium. This is not an exotic assumption; it is the typical pattern and it is easy to extend its results to cover cases with more fixed points. By the same token, it is also easy to simplify those same results if we want to consider a case with a unique equilibrium.[1]

Party Competition

We now turn to the game's first stage, the selection of platforms τ_L and τ_R. The fundamental results on the spatial models of voting have shown that when parties have ideological preferences and there is electoral uncertainty, the Downsian convergence to the median voter no longer holds (Roemer 2001b). With universal turnout, this often implies that if we want to study models with policy divergence across parties we need to introduce exogenous sources of uncertainty. In contrast, in the present model uncertainty emerges endogenously as a result of the multiple equilibria of the voting game played by the citizens. To see more clearly the implications of this, let's start by specifying the payoff function of the parties.

We began our discussion of political parties in this model by assuming that they do have ideological preference above and beyond their purely office-seeking motives. To capture this, we have been assuming that the parties represent the preferences of distinct income levels, so that L's payoff coincides with that of an agent with income y_L and R's payoff with that of an agent with income y_R and that $y_L < y_R$ so that L is, so to speak, the party of the "poor." I do not choose this setting out of a belief in its objective accuracy; although many democracies have party systems that seem to satisfy this description, in general it may not be wise to take that as an axiom. But since our interest here is in the patterns of correlation between income and voting, this framework is at once simple and reasonable.

In choosing their strategies, the parties must take into account the way their own platforms affect the ultimate policy outcome. Just running on a platform that coincides with their ideological ideal point may be suboptimal if that leads to electoral disaster and, hence, to giving enormous decision-making power to the opponent.

If we were to assume that $c_k = 0$ for every citizen, we would be back to the standard case of costless voting and universal turnout. With every citizen voting, the

[1] Although a full discussion would take us too far from our central topic, the reader must keep in mind that generically games have an odd number of equilibria. This explains why I do not consider the case of two fixed points.

electoral fortunes of each party are simply determined by the distribution of preferences among the populace; the model would have no electoral uncertainty unless we introduce some additional, exogenous mechanism that keeps some citizens from voting. For our purposes this would not work because such a mechanism would not establish a connection between electoral participation and the citizens' preferences as informed by their objective characteristics.

Instead, in the model we are analyzing the cost of voting is positive for at least some voters so that turnout is endogenously determined in equilibrium. If the game has multiple equilibria, this means that the election's outcome is inherently uncertain even if we consider τ_L and τ_R as fixed.

Facing this uncertainty, the parties need to form probability estimates for each equilibrium, something that can be obtained from the stability analysis of the previous chapter. Under the assumptions introduced, all the profiles of initial belief conditions m^0 such that $m^0 < m^u$ belong to the stability set of equilibrium $\underline{m}$ and, conversely, all those profiles for which $m^0 > m^u$ form the stability set of $\bar{m}$. By the same token, we know that the equilibrium m^u has no stability set of its own other than itself. Letting the function π represent the probability distribution of initial belief conditions, we can put all this together and form the parties' expected payoff function W_L and W_R:

$$\begin{aligned}
W_L(\tau_L, \tau_R) &= (V_L(\tau_L, y_L)g(\underline{m}) + V_L(\tau_R, y_L)(1 - g(\underline{m})))\pi(m^u) + \\
&\quad (V_L(\tau_L, y_L)g(\bar{m}) + V_L(\tau_R, y_L)(1 - g(\bar{m})))(1 - \pi(m^u)), \\
W_R(\tau_L, \tau_R) &= (V_R(\tau_L, y_R)g(\underline{m}) + V_R(\tau_R, y_L)(1 - g(\underline{m})))\pi(m^u) + \\
&\quad (V_R(\tau_L, y_R)g(\bar{m}) + V_R(\tau_R, y_R)(1 - g(\bar{m})))(1 - \pi(m^u)).
\end{aligned}$$

So, to complete the backward induction analysis of the game we need to find the Nash equilibrium of this first stage, that is, we need to find the values τ_L^* and τ_R^* that maximize these two functions simultaneously. At this level of generality we cannot calculate an exact solution, but we know enough about the structure of the game to prove some important propositions about it. Before, we shall notice some basic properties of the equilibrium platforms.

Lemma 1 *Let τ_L^* and τ_R^* be the equilibrium strategies of L and R respectively. Then: $\tau_L^* > \tau_R^*$.*

Proof The first statement can be proven by noticing that, if $\tau_L^* \geq \tau_R^*$, then each party could benefit from a unilateral deviation by choosing a platform closer to its ideal point: no matter how unlikely victory may be in that case, this deviation will tilt the expected outcome toward the party in question.

This has an important consequence for the analysis: the δ_i functions of some of the supporters of L have a different shape from the δ_j functions of the supporters of R. For any given tax rate, there is an income level below which the agent's optimum is at the corner solution. This means that, a pair of platforms $\tau_L^* > \tau_R^*$ partitions the electorate into three types. First, there are those citizens for whom $y_k > \alpha h(\tau_L \bar{y})/(1-\alpha)(1-\tau_L)$. These citizens, who will be called Type I citizens, have their consumer optimum in the linear portion of the budget constraint regardless of the tax rate implemented. Type II citizens are those such that $\alpha h(\tau_R \bar{y})/(1-\alpha)(1-\tau_R) < y_k < \alpha h(\tau_L \bar{y})/(1-\alpha)(1-\tau_L)$. For them, their consumer optimum under τ_L is at the corner solution but is in the linear portion under τ_R. Finally, Type III citizens have income $y_k < \alpha h(\tau_R \bar{y})/(1-\alpha)(1-\tau_R)$ so that under both tax rates their consumer optimum is at the corner solution. Since $\tau_L^* > \tau_R^*$ this partition is such that all the supporters of R are Type I citizens whereas the supporters of L may belong to any of the three types. The function $\delta_i(\tau_L, \tau_R)$ is:

- $\log(y_i(1-\tau_L) + h(\tau_L \bar{y})) - \log(y_i(1-\tau_R) + h(\tau_R \bar{y}))$ if i is Type I,
- $\alpha \log(y_i(1-\tau_L)) + (1-\alpha)\log h(\tau_L \bar{y}) - \log(y_i(1-\tau)R + h(\tau_R \bar{y}))$ if i is Type II and
- $\alpha(\log(1-\tau_L) - \log(1-\tau_R)) + (1-\alpha)(\log h(\tau_L \bar{y}) - \log h(\tau_R \bar{y}))$ if i is Type III.

Comparative Statics

Let's now use the model to address the central question of this chapter, that is, whether the politico-economic structure of redistribution plays a role in shaping the patterns of electoral turnout. There are two parameters of interest for our purposes: α and y. The first parameter tells us how big is the share of the government-provided goods in the overall consumption budget of the individuals. The larger α, the larger the role of privately-provided goods, that is, the smaller the role of redistribution in changing the consumers' utility. The logic of the argument introduced at the beginning of this chapter indicates that, as α grows large, the effect on the recipients of redistribution of an increase in the tax rate becomes smaller because the goods they actually receive play a smaller role in their overall consumption.

The parameter y will allow us to study the effects on turnout of the underlying income distribution. Intuitively, higher degrees of inequality lead to higher demand for redistribution because there is a larger set of potential beneficiaries from it, so that it becomes less costly in electoral terms for the party representing low-income voters to propose increased tax rates. This is, in fact, the standard argument since the classical Meltzer-Richard model. Our model captures these dynamics as well which, when taken together with the preceding argument, helps us refine our analysis of socio-economic bias in turnout.

The following theorem is the basis of our comparative statics results. First I will state it formally and leave the conceptual discussion for a later, non-technical section.

Theorem 5 *The equilibrium platforms τ_L^* and τ_R^* satisfy the following properties:*

- $\frac{\partial \tau_L^*}{\partial \alpha} < 0$, $\frac{\partial \tau_L^*}{\partial \alpha} > 0$ *and*
- *Let $\vec{y} = (y_1, \ldots, y_N)$ be an income profile of the economic environment and $\vec{y}'$ another income profile identical to $\vec{y}$ except for components k_1 and k_2. Suppose $y_{k_1} < y_{k_2}$ and $y'_{k_1} = y_{k_1} - \epsilon$, $y'_{k_2} = y_{k_2} + \epsilon$. Then, $\tau_L^*(\vec{y}) \leq \tau_L^*(\vec{y}')$ and $\tau_R^*(\vec{y}) \leq \tau_R^*(\vec{y}')$.*

Remark The theorem says that increases in the share of private goods vs. government provided goods shift the political equilibrium toward lower taxes whereas increases in income inequality (defined as a transfer from a poorer individual to a richer one) shift the equilibrium toward higher taxes.

Proof Both statements follow from an application of monotone comparative statics (Milgrom and Shannon 1994). To prove the first statement, let's compare the indirect utility function of Type I citizen with that of Types II and III.

Consider an arbitrary Type I citizen with income $y_i < y^0$:

$$\frac{\partial^2 \delta_i}{\partial \tau_L \partial \alpha} = 0.$$

Instead, if i is a Type II or III citizen, then:

$$\frac{\partial^2 \delta_i}{\partial \tau_L \partial \alpha} = -\frac{1}{1-\tau_L} - (1-\alpha)\frac{h'(\tau_L \bar{y})\bar{y}}{h(\tau_L \bar{y})} < 0.$$

Since all the supporters of party R are Type I whereas those of party L belong to the three types, this means that, when we aggregate δ_i and δ_j across income levels:

$$\frac{\partial^2 \gamma_R}{\partial \tau_R \partial \alpha} = 0,$$

$$\frac{\partial^2 \gamma_L}{\partial \tau_L \partial \alpha} < 0.$$

The equilibrium vote for L, V_L^* is determined as the solution to the equation $V_L - \gamma_L(V_L, \alpha)$. Applying the implicit function theorem to this equation we establish that:

$$\frac{dV_L^*}{d\tau_L} = \frac{\frac{\partial \gamma_L}{\partial \tau_L}}{1 - \frac{\partial \gamma_L}{\partial V_L}}.$$

We are interested in the magnitude of the changes in the different equilibria of the game. Since they respond in opposite directions to the parameters, let's focus on their absolute values. Differentiating again with respect to α we conclude that:

$$\frac{d}{d\alpha}\left|\frac{dV_L}{d\tau_L}\right| = \frac{\partial^2 \gamma_L}{\partial \tau_L \partial \alpha}\left|1 - \frac{\partial \gamma_L}{\partial V_L}\right|^{-1} < 0.$$

Since $m = V_L^* - V_R^*$, then m inherits the same signs of partial derivatives.

The preceding reasoning implies that m has non-increasing differences in (τ_L, α) so that $\tau_L^+(\alpha) = \arg\max_{\tau_L} m(\tau_L, \alpha)$ is non-increasing in α. An important property of monotone comparative statics is that the non-increasing character of the optimizer is preserved under monotonically increasing functions so, if we redefine $\tau_L^+(\alpha) = \arg\max_\tau h(m(\tau_L, \alpha)$ where h is monotonically increasing, then $\tau_L^+(\alpha)$ is still non-increasing in α. Party L's payoff function W_L is:

$$W_L(\tau_L) = \delta_L(\tau_L, y_L)[g(\bar{m}(\tau_L)) - (g(\bar{m}(\tau_L)) - g(\underline{m}(\tau_L)))\pi(m^u(\tau_l))] + V_L(\tau_R, y_L).$$

This function is monotonically increasing in all the variables $\bar{m}, m^u, \underline{m}$. Furthermore, the preceding analysis shows that $\delta_L(\tau_L, y_L)$, which is a special case of δ_i, also has non-increasing differences in (τ_L, α). Therefore, $\tau_L^*(\alpha) = \arg\max_\tau W(\tau_L, y_L|\alpha)$ is non-increasing in α.

The second statement of the theorem can be proven using an identical procedure noting that for any k,

$$\frac{\partial^2 \delta_k}{\partial \tau_L \partial y_k} = 0$$

if k is a Type II or Type III citizen but that

$$\frac{\partial^2 \delta_k}{\partial \tau_L \partial y_k} < 0$$

if k is a Type I citizen. Therefore, the analysis made for the first statement goes through with the signs of change inverted.

4.2 A Preliminary Discussion of the Results

By some standard, the preceding analysis might seem laborious so it is now appropriate to assess its payoff. I contend that the preceding model makes two contributions to our understanding of turnout, a general one and a particular one.

The general contribution is that it is one example of the type of models that can be developed by virtue of the advances we have made in the preceding chapters. Once we go beyond the turnout paradox it becomes possible to study how electoral participation responds to structural changes that affect the citizen's benefits and costs of voting. In this chapter I have illustrated this by offering a model in which the economic environment and the specifics of the mechanisms of political redistribution interact to affect such costs and benefits. The result is a set of testable hypotheses that relate the observable patterns of turnout with deeper economic forces. Furthermore, this approach to the problem allows us to bring together two strands of research that have remained separate in the rational-choice tradition: the spatial models of electoral competition and the models of turnout. The literature on spatial models has consistently assumed universal turnout, so as not to deal with the paradox of voting. Reciprocally, most models of turnout have tended to bracket the role of parties and policy platforms in shaping participation. But it is fair to say that the literature on turnout has largely been unable to incorporate ideas from models of spatial competition. The model I just presented brings these two approaches together by offering a "general equilibrium" analysis of turnout where the citizens' decisions to vote and the parties' choices of platforms influence each other.

It was not my intention to present a complete model of the determinants of turnout in a market economy but an illustration of how to build such models. To that extent, it may be premature to subject the current model to a test and to pronounce it a success or a failure. But neither is the model so preliminary and tentative that it cannot offer some insights into a substantive discussion of turnout so let's study its main propositions at length.

In a nutshell, the model argues that the politico-economic structure of redistribution in contemporary economies plays a role in shaping the patterns of turnout we observe. The mechanism through which this happens operates at the microeconomic level: the fact that redistribution is targeted to some specific goods creates an asymmetry in terms of the consumer options that both payers and beneficiaries of the welfare state face.

This asymmetry means that, whereas the payers perceive an increase in redistribution as a dollar-for-dollar reduction of their income, the same increase, when seen from the beneficiaries' point of view, translates into an increase of only those goods transferred by the government, goods subject to marginal diminishing returns. All else equal, the smaller the share of those goods in the total consumption basket of individuals, the smaller the benefit from redistribution and, hence, the smaller the incentive a citizen has to vote for it. This presents the electoral defenders of redistribution with a problem: the increases in taxes that would be needed to encourage the beneficiaries of redistribution to vote alienate sizable segments of the rest of the electorate. So, if this trade-off is too steep, as it is if redistribution is too heavily concentrated on a few goods, the pro-redistribution party ends up, in equilibrium, adopting a more moderate stance.[2]

Not all else is equal, of course. Received wisdom in the study of redistribution claims that in highly unequal societies, the poorest citizens stand to gain much from any type of redistribution and therefore will have more incentives, not less, to vote for it. Theorem 5 tries to capture this tension. Plainly, it states that both arguments are correct so that the net effect depends on the specifics. Redistribution schemes that target too narrowly into goods with a small share in the total consumption basket depress the turnout of the recipients, thus inducing a class bias in electoral participation. But increases in inequality make the poor vote more, counteracting this bias.

---

[2]The central role of trade-offs across constituencies places this model in close proximity to the classic analysis of the dilemma of left-wing parties in Europe by Przeworski and Sprague (1986). The clearest difference is, of course, that in their analysis the turnout decisions of the citizens were not explicitly modeled.

At the current stage, these hypotheses are difficult to test because cross-national data on class bias in turnout is in short supply. But there are some elements that might serve as circumstantial evidence. It is well-known that the correlation between turnout and income is higher in the US than in Europe (Lijphart 1997) and, at least as a first approximation, it could be argued that the American welfare state covers a smaller portion of the individuals' overall consumption basket. Given that the preceding analysis assigns a significant role to differences in income distribution and the structure of the welfare state, it is regrettable that most of the available studies of socioeconomic bias in electoral participation focus on developed economies. A true comparative test would require looking at Third World countries. Still, some data suggest that the socioeconomic bias in the US is even higher than the (non-negligible) bias in some Third World countries, such as those of Central America (Seligson et al. 1995). The reasoning I have presented implies that, the more the benefits from redistribution resemble a pure cash transfer, as would be the case in a "basic income program," the smaller the discrepancies in voting between payers and beneficiaries. Interestingly, within the US, the country where patterns of turnout have been studied at most length, there is one distinct set of citizens that receive cash transfers and happen to vote at higher rates than the rest: pensioners. I would be tempted to offer this as further evidence of my argument if it were not for the fact that there are many alternative explanations for the age-gap in voting in the US and at this point it is hard to discern which is correct.

Given the dearth of empirical studies, it would be foolhardy to make too much of these findings. But if they hold up, the pattern that emerges is consistent with the preceding discussion. In light of this, a case could be made that the socioeconomic bias in voting in the US is yet another case of "American exceptionalism." Certainly, when compared to other advanced democracies, the US stands apart in this respect. But probably we can do better in trying to explain this exceptional case. The analysis of this chapter suggests that, against an ideal benchmark of full equality in voting, the US has the worst of two worlds: its mechanisms of redistribution are more narrowly focused on specific goods than those of European welfare states something that depresses turnout and, compounding matters, its low-income segment is, by world standards, relatively affluent. In this regard, it contrasts with the plight of poor citizens in Third World countries for whom their very poverty constitutes an incentive to vote in exchange for paltry transfers that would not bring to the polls the poorest citizens of a developed country.

The preceding remarks probably raise more questions than they answer. As I have repeatedly stated, my goal is to develop structural models of turnout, of which the one in this chapter is an example. Describing a model as structural is not an innocent label but one that carries several methodological and substantive implications. Therefore, I prefer to postpone a full-blown discussion to the next chapter, where it will not have to coexist uneasily with the technical details that abound in this one.

Chapter 5
Toward a Structural Theory of Turnout

Political scientists in need of reassurance about the worthiness of their endeavor can do worse than visualize Buenos Aires's Avenida 9 de Julio. One of the world's widest urban thoroughfares, by the time it reaches the obelisk, a major symbol of the city, it opens creating a large expanse that is often pressed into service as a public square. Reportedly, the largest demonstrations this open space has witnessed in recent memory were the closing acts of the presidential campaign of 1983, a feat that becomes more impressive when we consider that Argentina has won the football (soccer) World Cup twice. But these elections were special. A military regime that "disappeared" more than 20,000 citizens, after exiling, imprisoning and torturing many others, that had brought about a serious economic collapse and that, in its agony, had embarked upon an irresponsible international war which it fought incompetently, was about to exit history through the back door.

The exhilaration Argentineans felt at that point is hardly unique and we can find similar examples in the democratic transitions that many other countries have gone through. But a richer, more complex picture and a different set of reflections, emerge when we go back in history 7 years. March 24, 1976 marks the beginning of this same dictatorship, when Argentina's top brass detained Isabel Martínez de Perón, up to that point the country's constitutional president. The democratic fervor that would erupt 7 years later was nowhere to be seen back then: no major demonstration, no Molotov cocktails hurled at the Army, no heroic death in defense of the government.

Certainly, it is not true that Argentineans simply did not appreciate democracy as a form of government by 1976. Three years before that, in 1973, they had greeted with unstoppable enthusiasm the newly elected government of Juan Domingo Perón (who died shortly after so that Isabel, his widow and vice-president took office). Their disdain of 1976 was not directed against an abstract principle of representation but against the practice of, to paraphrase the official term of Communist regimes, "really existing democracy." By the time of that fateful March 24 the government was, in essence, a political corpse in wait of burial. Even its democratic veneer was beginning to peel off given that it had already resorted to the harsh repressive tactics that would later become the successor junta's trademark.

© Springer International Publishing AG 2018
L. F. Medina Sierra, *Beyond the Turnout Paradox*, SpringerBriefs in Political Science,
https://doi.org/10.1007/978-3-319-73948-9_5

Beyond the discussion of a particular, if fascinating, political process, there are several lessons that we can take away from this example. Given the central role that democracy plays in contemporary political science, and the normative preeminence it enjoys in political theory, it is worth remembering that some democracies in real life collapse in ignominy with hardly anyone willing to shed tears (let alone blood) for them. Any theory about the causes and consequences of democracy must come to terms with this.

At times, our theoretical models of democracy seem designed to capture only the spirit of our 1983 example, slighting the one of 1976 at their peril. Much of the normative appeal of democratic rule is that in it, the only limits to the realm of possibilities is the consent of the governed. In a fully competitive democracy, so the argument goes, every political alternative is on the table for the polity to consider. In that view, elections represent a prudently regulated moment of contingency, when every institution and law is up for grabs, and when the only options that have their permanence guaranteed, that is, until the next election, are those consistent with the will of the majority. But, as Argentineans experienced starkly in 1976, sometimes democracy does not deliver on its promise of offering a free space where the citizenry can come together and chart whatever path it wants for itself. Sometimes democracy itself becomes hamstrung by the very institutions, laws, policies and practices it is supposed to preside over. When that happens, it becomes understandable why citizens turn their backs on the democratic process. On occasions, as in the example above, this might take the extreme form of simply shrugging at the sight of the tanks rolling down the streets. On others, much more frequent, much less dramatic, this might be accomplished by not voting.

With their emphasis on the role of political competition, rational-choice models are often good exponents of the tradition that emphasizes the open-ended possibilities of democratic rule. The bulk of the literature in the tradition of spatial models of voting draws on the Downsian approach where parties, driven by their desire to win office, offer different options to the voters until only the one that commands the majority's support is left standing. To be sure, not every spatial model results in convergence to the median.[1] But it remains the case that in those models all the alternatives on the policy space are created equal, even if their fate is not.

Although less explicit, partly as a result of the turnout paradox, the rational-choice models of electoral participation are also to some extent beholden to the view of democracy as a type of governance where individual decisions are the ultimate arbiter. When, for instance, rational-choice models of turnout emphasize the role of information in determining who votes and who does not, they are implicitly holding fast to the notion that abstention is some kind of deviant case where the standards of free and full scrutiny of alternatives have not been met for reasons that have to be found outside of politics, be they differences in education or differences in the individual costs of becoming politically informed.[2]

[1] See, for instance, Roemer (2001b).

[2] See, for instance, the examples discussed in the previous chapter, especially Feddersen and Pesendorfer (1997) and Filer et al. (1993).

One of my central goals in this book has been to offer a corrective to this bias with my insistence on developing structural models of electoral participation. I do not think I am there yet. Moreover, I do not think we should go all the way there. "Corrective" is the instrumental word. I believe that a good theory of democracy must retain the element of agency emphasized by the traditions from which rational-choice theory has emerged.

If this were a book trying to put forward a normative theory of democracy, this caution would be rather optional. The reader could discount it as an affectation of an author too afraid to follow his instincts one way or another. But my goal is not to offer a notion of what democracy ought to be, but to develop tools to describe and understand the empirical patterns of democracies. Then, the question is not whether democracy "is" a sphere of boundless political possibilities or if it "is" an outgrowth of deeper economic and social realities and therefore limited in its potential for transformation: democracy is both to some extent. From an empirical point of view, differences of degree are what matters. Every democracy combines both elements, each one in its own way. These specificities are what a satisfactory theory of democracy should try to capture.

This book is based on two premises: (*a*) the rational-choice paradigm offers the best promise to advance in the direction I just sketched and (*b*) to that end, some changes in the paradigm are needed, but not a major overhaul. For reasons of space (and lack of a fireproof argument) I cannot defend (*a*) here. Instead, I will summarize the steps I have taken to deliver on (*b*).

5.1 Participation as Structured Choice

The single most important impediment to develop a structural analysis of electoral participation based on rational-choice theory is the turnout paradox. Once that roadblock is removed, only minor adjustments remain.

If the turnout paradox is always correct, if it afflicts any model of elections with a sizable citizenry, then electoral participation can only be explained through an appeal to psychological dispositions of the voters that have little or no connection to the social structure they inhabit. But as I have shown in Chap. 1, the scope of the paradox is not that large. On the contrary, once we evaluate the real extent of the damage done by the paradox, it turns out that it is possible to build many rational-choice models of voting with realistically high turnout. This is true even in the canonical models that originally gave rise to concerns about the paradox. Equilibrium high turnout is not an artifact of some exotic assumptions built into the model with the purpose of avoiding the paradox; it is a straightforward outcome of the original assumptions. Of course, that original model can be modified and improved and I discuss such modifications and improvements in Chap. 2, but the fact that the turnout paradox is not as serious as we often take it to be remains true even in the original setting.

I have paid special attention to one specific improvement because it helps us polish some technical rough edges: the continuous outcome function. The standard analysis of voting assumes that victory in an election is all-or-nothing and that a margin of just one vote is enough to secure it. In spite of being unrealistic, oversimplifying many crucial aspects of any political process, it is understandable that in the early stages of modeling elections we want to cover this case given that it is the simplest possible assumption. But, although simple to state, it leads to high-turnout equilibria that are unnecessarily baroque: equilibrium profiles are an unwieldy combination of pure and mixed strategies with proportions that need to be carefully characterized. The would-be gains in terms of simplicity, at the expense of realism, evaporate. Instead, a continuous outcome function results in high-turnout equilibria that are easy to describe and that jive with our everyday intuition: in an election, citizens with a cost-benefit ratio above an endogenous threshold abstain and those with a cost-benefit ratio below said threshold vote. The remaining problem is to calculate the value of this threshold but this is no harder than calculating the corresponding threshold in a coordination game.

For our purposes, the most important consequence of overcoming the turnout paradox is that it makes possible to develop models of turnout where the costs and benefits of voting are the major driving force. There are all sorts of costs and benefits but there is no question that some of them respond to social and economic phenomena.

There are many things that make voting costly. Sometimes it is simply the opportunity cost of taking time off to go to the polls, sometimes, a case widely analyzed in the US, there are registration laws that make it hard, others, the most extreme, would-be voters may risk intimidation and violence. Some of these costs are unrelated to systematic social processes, but others are not. I already mentioned briefly the many attempts, from poll taxes to actual killing of those involved in registration drives, that Southern segregationists used in the US to suppress the black vote. But the US is hardly the only democracy where this has happened. Similar phenomena, often much more lethal, occur in other places. In all these cases, the costs of voting are not randomly allocated across citizens but respond to very specific conditions in the polity.

In probably subtler ways, the same holds for the benefits of voting. In the calculus of voting the benefits represent the ideological evaluation that citizens make of the different alternatives on offer. As such, they could be merely subjective traits and no violence to the basic model would be done by treating them that way. But a view shared by many long-standing traditions in political science is that individuals do not simply chose their ideological perceptions detached from their everyday life.

In the models I have developed above, ideology results simply from the evaluation that individuals make of the socioeconomic environment they inhabit and the specific role they play in it. This is hardly a shocking view. To some, it is common to Marxist models of class struggle, much of the political sociology in the tradition of Lipset and Rokkan (1967) (e.g. Luebbert 1991) and rational-choice models of political economy. These three examples differ in the details and the definitions, but they all share some common kernel. In the case of the rational-choice paradigm, this evaluation is based on the rational assessment that individuals make of the way in which objective changes in the environment affect their own interests.

Just stating the view already suggests the many things that can be wrong with it. The reliance on rationality and self-interest is problematic and deserves careful scrutiny and debate. But this is not the place for it. Instead of substantiating this position in the abstract, a rather futile exercise, here I will work through some of its implications for our analysis. When the time comes for the full debate, these implications can become judgment elements.

5.2 Formal Rights and Material Imperatives

Rational-choice theories of politics have often been criticized for assuming that the individuals' preferences are exogenously given. Examples of this complaint can be found among historical institutionalists (e.g. Thelen and Steinmo 1992). Whatever the merits of this criticism, the analysis developed in this book takes steps toward a response. To understand why, let's study at length the model presented in Chap. 4.

In every modern democracy the State is involved, in varying degrees, in tasks that modify the allocation of resources that would result from a purely competitive market. Many of these tasks take the form of direct redistribution aimed at reducing the market-generated income inequality. This process has been a running theme through a vigorous literature in the rational-choice tradition.

Ever since the "workhorse" model of Meltzer and Richard (1981), the preferences of the agents in these models are inferred based on their location in the economic structure so that, for instance, agents with income below the median, who stand to benefit from a straightforward, redistributive tax would support it. Strictly speaking, these models assume that the citizens' preferences are exogenous. But that is not necessarily a bad thing. The question is whether a subtler treatment of preferences may illuminate some aspects that the simpler framework might overlook. In this case the answer is, I believe, "yes."

Once the turnout paradox stops being a concern for a game-theoretic model of voting, we can start focusing on how the broader structure of costs and benefits affects electoral participation and, more generally, collective policy-making. In the model I have presented, that structure contains two main components: the "before-government" income distribution, which is common to the standard models of the literature, and the policy-specific mechanisms of redistribution that are, instead, normally ignored in the analysis.

The Birdseye view of welfare states as mechanisms that tax and transfer income is inadequate for some purposes. Welfare states are more than an array of conveyor belts of income; they are highly complex institutions that encapsulate and reproduce key aspects of a social contract. For instance, welfare states affect the way families operate, including gender and intergenerational relations. This example, unrelated to our present object of analysis, ought to illustrate the extent to which the specifics of a welfare state are connected to aspects of social life that go beyond income.

One of the key aspects of a welfare state is the way in which it defines citizenship and rights. In a market economy, non-coercive, non-fraudulent transactions generate enforceable property rights. In the typical case, these property rights cover all the possible events: within some broad legal limits, the rightful owner of *X* can use it anyway she wants, can decide not to use it, can dispose of it, can give it to whoever she pleases, can trade it for whatever she wants and so on. The welfare state modifies this structure in two ways: it redefines part of a citizen's income as a tax obligation to the state, thus making it alienable, and it creates a restricted type of property rights for recipients of the transfers. The model I presented focuses on this second aspect.

When a welfare state transfers goods to a citizen, it is granting the individual certain rights over the enjoyment of said goods but those rights have a more limited scope than the one attached to rights over goods acquired in the market. Specifically, these goods cannot simply be traded away in exchange for money or other goods. Not every transfer fits this description. Pensions and other cash transfers with no strings attached are exceptions. But the overall principle is that the welfare state's transfers are subject to conditions that, in the case of goods, amount to a "not for resale" constraint.

The microeconomic analysis I presented above showed that this "not for resale" constraint affects the citizens' political preferences over redistribution. This asymmetry between recipients and payers of the welfare state translates into an asymmetry about the way they each evaluate the policy options on offer. In particular, the marginal benefits and losses of any redistributive policy differ for both categories of citizens. Whereas for the net payers, the marginal effect of an increase in taxes is a marginal loss of income, for the beneficiaries the corresponding gain is a marginal increase in the availability of specific goods.

In this sense, the model treats preferences as endogenous. True, the agents still have some immutable higher-order preferences for income, but it is hard to see what would be gained by trying to endogeneize these. Instead, their political preferences, the ones that drive their choices over the options they face in political competition, are indeed shaped by the very structure that defines those same options.[3]

Before we consider the role of countervailing forces, let's analyze the broader implications of this for political competition and electoral participation. This particular structure of the citizens' preferences means that, as they compete for votes, the parties face different trade-offs depending on the type of constituency they are trying to mobilize, something at variance with what tends to happen in standard models of spatial competition.

If the parties have ideological preferences, they need to choose their strategies mindful of the fact that policy proposals that may be ideologically satisfying to its strongest supporters may not be popular with voters in the center, which are, after all, also necessary. From a technical point of view, the equilibrium of a spatial

[3]When addressing the controversy over endogenous preferences, Clark (1998) makes a general argument similar in spirit to the one made here.

voting game is meant to capture this tension: each party chooses the strategy that makes the best out of this extreme-center trade-off, provided what the competing party is doing. A well-established principle of this type of analysis is that changes in the tradeoff change the equilibrium; in any decision problem with conflicting goals, such as the one at hand, the harder it is to obtain one of the goals compared to the others, the less resources will be devoted to it. We see this in the model at hand. The party that wishes to increase the size of redistribution faces a steeper trade-off than its opponent: to mobilize the poorest voters in the polity it would need to propose tax increases that would not compensate the loss in votes among the middle-income ones. As a result, in equilibrium this party proposes levels of redistribution that are more modest than what it would have done without the prevailing asymmetry between payers and recipients. In turn, this results in low levels of electoral participation among low-income citizens.

Given the source of this effect, how big it is depends on the type of transfers. The smaller the marginal utility of the goods transferred, the lower will be the gains that their recipients experience in case of an increase in the tax rates and, hence, the steeper will be trade-off of votes between low- and middle-income voters for any party trying to propose such increase.

The type of transfers is not the only determinant of their utility for consumers: income levels matter as well. The loss in utility an individual suffers as a result of a reduction in disposable income, in turn a result of higher taxes, is smaller the higher the income level. After all, income itself is also subject to diminishing marginal returns. Just as the "not for resale" constraint faced by recipients affects the willingness to vote of low-income citizens, at the other extreme of the income distribution affluence reduces the marginal salience of the tax, creating a similar problem for the party against redistribution.

If we apply the same logic as above to this case, we see that higher levels of income among the already high-income voters, that is, increases in inequality, make the low-taxes party face a steeper trade-off between the extreme and the center. The result is that, in equilibrium, this party's stance becomes more moderate.

The main result of Chap. 4 brings these two effects together in a mathematical statement about the comparative statics of the political equilibrium. They point in opposite directions: highly unequal income distributions depress turnout among high-income voters and tilt the political equilibrium toward higher taxes, narrowly focused welfare states depress turnout among low-income voters and tilt the political equilibrium toward lower taxes. The first effect should be familiar from the standard models of electoral redistribution: inequality increases demands for redistribution.[4] Instead, the second effect can only be brought out if we extend the analysis to cover redistributive models with several goods: the model presented here is a generalization of the one-good models well-known in the literature. As such, its results offer a more nuanced set of hypotheses.

[4]Apart from the example already adduced of Meltzer and Richard (1981), we find the same logic at work in the models of Acemoglu and Robinson (2006) and Boix (2003).

Although limited, this model illustrates the programmatic point I have stressed throughout the book: with the turnout paradox no longer a hindrance, it becomes possible to study how deep structural forces shape the observed patterns of electoral turnout. In this case, we have not made any assumption about ad hoc asymmetries across participation costs for voters. The underlying income distribution and the structure of the welfare state are the ones responsible for the class bias.

From a broader perspective, it may be argued that the structure of the welfare state is itself the outcome of a political equilibrium. This is undeniable and a more general model should consider the possibility that the parties compete not just by proposing different levels of taxation but also different mechanisms of redistribution. But while this is a limitation of the model, it is not devastating namely for two reasons.

First, as already mentioned, the mechanisms favored by any given welfare state reflect deeply entrenched principles of the social pact. The type of property rights established by the welfare state, the extent to which income from transfers is treated differently from market income is hardly the fickle outcome of some electoral cycle. To the contrary, the specifics of the consensus a society reaches in these matters come into being through lengthy political battles. Conversely, this same consensus is often shattered only after large scale social transformations.

Second, although the model left unspecified the determinants of the mechanisms of redistribution, it shows us ways in which such mechanisms gain stability in the political process. One decisive advantage of the models of turnout I have discussed is that they allow us to treat electoral participation as part of a broader equilibrium among several forces. In this model the pattern of turnout is determined jointly with the equilibrium of the spatial competition among parties. As a result, welfare states that place tight restrictions on the type of goods they transfer will, all else equal, generate political equilibria with lower levels of taxation which will, in turn, make it difficult to fund further expansions. Conversely, welfare states that cover more goods, services and risks, thus coming closer to pure cash transfers, will, again, all else equal, support equilibria with higher levels of taxation that will make them politically self-sustaining.

5.3 Inequality and Representation

While the class bias in turnout is something of a well-established regularity, it is not clear what it means. The interpretation that, so to speak, suggests itself is that class bias undermines democracy. If citizens with a specific type of preferences are consistently excluded from the democratic process, even if that exclusion is voluntary, then a case can be made that the outcome does not truly do what they are supposed to: treat equally all the different ideological viewpoints.

But some empirical findings suggest that the case may not be so simple. The claim that socioeconomic bias in turnout is itself a disfunction of representation assumes that it is possible to map a citizen's station in life into her ideological views.

In principle, this should be testable and a large literature has done exactly that. (See, for instance, Lutz and Marsh (2007) and the other articles in the symposium.) Recent findings suggest that the impact of turnout on elections is rather negligible and unsystematic. The conventional view that decreases in electoral participation hurt left-of-center parties seems to be disproved by the fact that usually there are no major ideological differences between voters and non-voters. In fact, there are often cases when nonvoters are, on average, to the right of voters so that in those instances, if anything, declines in turnout hurt the parties on the right.

Without stepping into the empirical controversy between these two arguments, the analysis we have made so far suggests some points that are worth considering. A central theme of the preceding discussion is that the citizens' preferences are not a detached evaluation of political alternatives but the outcome of a process where different elements of the social, economic and political environment interact. As our model of electoral redistribution shows, the electoral behavior of individuals is shaped jointly with the political choices of the competing political parties against the backdrop of a complex structure that includes economic realities, such as the income distribution, and long-standing aspects of the social pact, such as the mechanisms through which the welfare state is expected to work. It is not possible to understand the citizens' choices without understanding the alternatives that political parties offer them and, reciprocally, those alternatives are chosen by the parties taking into account the electorate's preferences.

Taking all this into account, it is possible to offer a different interpretation of the alleged similarity of preferences between voters and nonvoters. To fix ideas, imagine that the position of one party is given, for instance, the party of the right, and let's see what happens as the other converges toward it. If the discrepancy between the two parties is large, the average benefits of voting are high and so high numbers of citizens vote. With low levels of ideological distance between the parties, the average benefits of voting are low and so turnout is low. But this has another consequence: since the level of benefits is low among voters, this also means that voters are now more similar to nonvoters in this regard.

Applying this reasoning to our model of electoral participation in a welfare state, we conclude that polities with high class bias are also polities where the two parties tend to converge around policy platforms of low redistribution. In turn, this may reduce the observed ideological differences between voters and nonvoters.

Thus, the seemingly clear normative question posed by low and biased turnout comes under a different light. In the analysis I have carried out here, these patterns of turnout are not a malady of democratic procedures but rather a manifestation of deeper structural factors having to do with the economic environment, the prevailing definition of citizens' property rights and the makeup of political parties. Whether low and biased turnout is unfair depends on our views of these underlying factors. If, for the sake of argument, we believe that the differential treatment between income from transfers and market income is unfair and that, say, the laws governing the financing of parties force them to moderate their views too much, then we could hardly conclude that the fact that voters and nonvoters have similar preferences render these issues moot.

This is a concrete example of the broader concern with which I opened this discussion. Once we consider that democratic procedures are embedded in a broader structure, we must recognize that even though democracy implies open political competition, this does not mean that through democratic procedures the citizenry can simply start from scratch the task of designing its own laws, institutions and policies. Instead, those same laws, institutions and policies will somehow reflect the power relations that define the economy and the social order.

5.4 Beyond Determinism

While the rational-choice paradigm could benefit from giving more salience to the way in which socioeconomic structures shape the workings of democratic procedures, it would be a mistake to take this to the extreme of reducing politics to a simple appendage of those structures. Elections cannot bring about complete overhauls of the polity, but neither are they exercises in irrelevant posturing. Some of the largest transformations in society over the twentieth century occurred as a result of elections. Such is the case, for instance, of many of the expansions and contractions of the welfare state in advanced democracies.

The enthusiasm that often explodes among the winners in a typical election night is a testament to the feeling of open-ended possibility that takes hold of the participants. We could cynically dismiss that as collective self-deception but would miss an important point: at the end of the day, even if all the structural conditions point in one same direction, in a democracy it is up to the citizenry to act so that no outcome is truly inevitable.

One of the main strengths of rational-choice theory is precisely that, through its emphasis on the role of agency, it allows us to turn this insight into a rigorous and operational analytical tool. Nowhere is this more clear than in the case of multiple equilibria that forms the core of Chap. 3.

In games with multiple equilibria, the payoff structure is not enough to determine the outcome. Ultimately, the choices of individuals depend also on a set of mutual understandings and shared beliefs that allow each agent to form a judgment of the other players' behavior. Although it is impossible to determine a set of necessary conditions for multiple equilibria, they are a common property of games that require coordination across players. In that sense, voting games are not the exception.

This might seem to contradict all my preceding arguments. I have been insisting on the fact that election outcomes reflect deep structural forces at work but now, in recognizing that elections have multiple equilibria, I am saying that, whatever those forces dictate, the final outcome is contingent. But this contradiction only exists if we are determined to take an extreme view both on the role of structural factors and on contingency. Otherwise, both views can be reconciled with an appeal to the method of stability sets.

In game theory, an equilibrium results from a specific confluence of patterns of individual behavior that become mutually reinforcing. Each player chooses a strategy because it is best provided what the others are doing. If the game has multiple equilibria, this means that their decisions are buttressed by belief conditions that do not pertain to the payoffs. This does not mean, however, that every equilibrium is equally likely: some equilibria require very stringent belief conditions to emerge, others instead can result from a wider variety of such beliefs. Intuitively, it stands to reason that, when this is the case, some equilibria will be more likely than others. This is the intuition that the method of stability sets formalizes.

The keystone of this analysis is that the size of a an equilibrium's stability set, that is, the range of belief conditions that will give rise to it and that, hence, determine its likelihood, is in turn a function of the payoff structure. In a game with multiple equilibria, changes in the payoffs change the relative probability of each equilibrium.

Thus the two strands merge. When we analyze games with multiple equilibria, such as voting, thanks to the method of stability sets we can understand that, while the structural changes play a role in privileging some equilibria over others, by making them more likely, those same changes do not exhaust the role of the mutual beliefs and understandings that the players bring with them to the game. The models I have presented here exemplify this. In all of them, the structural parameters give us valuable information about the direction in which the outcome moves but they are never enough to determine it with certainty. In the context of elections, this means that other factors not related to the underlying payoffs need to be considered, factors that help or impede the voters' task of coordinating around certain choices, factors such as organization, culture, traditions and even charisma.

In my analysis of these models I have not made an attempt to introduce these variables systematically. This was a conscious choice driven by two considerations: first, I think that scholars in the rational-choice paradigm are already attuned to this problem so that the structural aspects were the ones in more need of emphasis, second, while I believe that the game-theoretic methods I have used here are perfectly capable of incorporating how the belief conditions come into play, I have doubts that the origins of those beliefs are, themselves, game-theoretic. If I am right, that will be yet another avenue of research to broaden the perspectives of rational-choice theory.

Although a source of interesting results, the turnout paradox was, at the end of the day, a puzzle of rational-choice's own making. It is, then, understandable that scholars that do not share this specific agenda have remained unimpressed by the paradox and indifferent to the attempts at solving it. I am convinced that rational choice theory has much to offer to other approaches, but also much to learn from them. So, an enriching dialog between traditions is a necessity. I hope that, with the turnout paradox reduced to its proper, peripheral role, such dialog will also become a reality.

Bibliography

Acemoglu, Daron and James A. Robinson. 2006. *Economic Origins of Dictatorship and Democracy*. New York: Cambridge University Press.

Ambrus, Attila. 2006. "Coalitional Rationalizability." *The Quarterly Journal of Economics* 121(3):pp. 903–929.

Anderson, Christopher J. and Yuliya V. Tverdova. 2001. "Winners, Losers, and Attitudes about Government in Contemporary Democracies." *International Political Science Review* 22(4):321–338.

Aumann, Robert. 1959. Acceptable Points in General Cooperative n-Person Games. In *Contributions to the Theory of Games IV*, ed. Harold W. Kuhn and Robert Duncan Luce. Princeton: Princeton University Press pp. 287–324.

Bacharach, Michael. 2006. *Beyond Individual Choice: Teams and Frames in Game Theory*. Princeton: Princeton University Press.

Bendor, Jonathan, Daniel Diermeier, David A. Siegel and Michael M. Ting. 2011. *A Behavioral Theory of Elections*. Princeton: Princeton University Press.

Bendor, Jonathan, Daniel Diermeier and Michael Ting. 2003. "A Behavioral Model of Turnout." *The American Political Science Review* 97(2):pp. 261–280.

Bernheim, B Douglas, Bezalel Peleg and Michael D Whinston. 1987. "Coalition Proof Nash Equilibria I: Concepts." *Journal of Economic Theory* (42):pp. 1–12.

Blais, André and Agnieszka Dobrzynska. 1998. "Turnout in electoral democracies." *European Journal of Political Research* 33(2):239–261.

Blais, André and R. K. Carty. 1990. "Does proportional representation foster voter turnout?" *European Journal of Political Research* 18(2):167–181.

Boix, Carles. 2003. *Democracy and Redistribution*. Cambridge: Cambridge University Press.

Burnham, Walter Dean. 1982. *The Current Crisis in American Politics*. New York: Oxford University Press.

Castanheira, Micael. 2003. "Victory margins and the paradox of voting." *European Journal of Political Economy* 19(4):817–841.

Clark, William Roberts. 1998. "Agents and Structures: Two Views of Preferences, Two Views of Institutions." *International Studies Quarterly* 42(2):245–270.

Cox, Gary W. 1988. "Closeness and Turnout: A Methodological Note." *Journal of Politics* 50(3):768–775.

Downs, Anthony. 1957. *An Economic Theory of Democracy*. New York: Harper.

Feddersen, Timothy J. and Alvaro Sandroni. 2006. "A Theory of Participation in Elections." *American Economic Review* 96(4):1271–1282.

© Springer International Publishing AG 2018

L. F. Medina Sierra, *Beyond the Turnout Paradox*, SpringerBriefs in Political Science,
https://doi.org/10.1007/978-3-319-73948-9

Feddersen, Timothy J. and Wolfgang Pesendorfer. 1997. "Voting Behavior and Information Aggregation in Elections with Private Information." *Econometrica* 65(5):1029–1058.

Feddersen, Timothy J. and Wolfgang Pesendorfer. 1999. "Abstention in Elections with Asymmetric Information and Diverse Preferences." *The American Political Science Review* 93(2):381–398.

Feddersen, Timothy, Sean Gailmard and Alvaro Sandroni. 2009. "Moral Bias in Large Elections: Theory and Experimental Evidence." *The American Political Science Review* 103(2):pp. 175–192.

Ferejohn, John A. and Morris P. Fiorina. 1974. "The Paradox of Not Voting: A Decision Theoretic Analysis." *American Political Science Review* 68(2):525–536.

Filer, John E., Lawrence W. Kenny and Rebecca B. Morton. 1993. "Redistribution, Income and Voting." *American Journal of Political Science* 37(1):63–87.

Fowler, James H. 2006. "Altruism and Turnout." *The Journal of Politics* 68(3):pp. 674–683.

Gerber, Alan S., Donald P. Green and Christopher W. Larimer. 2008. "Social Pressure and Voter Turnout: Evidence from a Large-Scale Field Experiment." *The American Political Science Review* 102(1):pp. 33–48.

Geys, Benny. 2006. "Explaining voter turnout: A review of aggregate-level research." *Electoral Studies* 25(4):637–663.

Grofman, Bernard and Guillermo Owen. 1984. "To Vote or Not to Vote: the Paradox of Nonvoting." *Public Choice* 42(3):311–325.

Harsanyi, John and Reinhardt Selten. 1988. *A General Theory of Equilibrium Selection.* Cambridge, MA: MIT Press.

Jackson, Matthew O. and Leeat Yariv. 2007. "Diffusion of Behavior and Equilibrium Properties in Network Games." *The American Economic Review* 97(2):pp. 92–98.

Kanazawa, Satoshi. 1998. "A Possible Solution to the Paradox of Voter Turnout." *Journal of Politics* 60(4):974–995.

Ledyard, John. 1984. "The Pure Theory of Two Candidate Elections." *Public Choice* 44:7–41.

Levine, David K. and Thomas R. Palfrey. 2007. "The Paradox of Voter Participation? A Laboratory Study." *American Political Science Review* 101(1):143–158.

Lijphart, Arend. 1997. "Unequal Participation: Democracy's Unresolved Dilemma." *American Political Science Review* 91(1):1–14.

Lipset, Seymour Martin. 1960. *Political Man: The Social Bases of Politics.* Garden City, NY: Doubleday.

Lipset, Seymour Martin and Stein Rokkan. 1967. Cleavage Structures, Party Systems and Voter Alignments: An Introduction. In *Party Systems and Voter Alignments: Cross-National Perspectives*, ed. Seymour Martin Lipset and Stein Rokkan. New York: Free Press pp. 1–64.

Luebbert, Greg. 1991. *Liberalism, Fascism or Social Democracy: Social Classes and the Political Origins of Regimes in Interwar Europe*. New York: Oxford University Press.

Lutz, Georg and Michael Marsh. 2007. "Introduction: Consequences of low turnout." *Electoral Studies* 26(3):539–547.

Medina, Luis Fernando. 2005. "The Comparative Statics of Collective Action: A Pragmatic Approach to Games with Multiple Equilibria." *Rationality and Society* 17(4):423–452.

Medina, Luis Fernando. 2007. *A Unified Theory of Collective Action and Social Change.* Ann Arbor, MI: University of Michigan Press.

Medina, Luis Fernando. 2011. "Is There Really a Turnout Paradox?" *Journal of Theoretical Politics* 23(3):382–399.

Medina, Luis Fernando. 2013. "The Analytical Foundations of Collective Action Theory: A Survey of Some Recent Results." *Annual Review of Political Science* (16):259–83.

Mehta, Judith, Chris Starmer and Robert Sugden. 1994. "The Nature of Salience: An Experimental Investigation of Pure Coordination Games." *The American Economic Review* 84(3):pp. 658–673.

Meltzer, Allan H. and Scott F. Richard. 1981. "A Rational Theory of the Size of Government." *Journal of Political Economy* 89(5):914–927.

Milgrom, Paul and Chris Shannon. 1994. "Monotone Comparative Statics." *Econometrica* 62(1):157–180.

Morton, Rebecca B. 1991. "Groups in Rational Turnout Models." *American Journal of Political Science* 35(3):758–776.

Myerson, Roger. 1991. *Game Theory: Analysis of Conflict*. Cambridge, MA: Harvard University Press.

Osborne, Martin and Ariel Rubinstein. 1994. *A Course in Game Theory*. MIT Press.

Palfrey, Thomas R. and Howard Rosenthal. 1983. "A Strategic Calculus of Voting." *Public Choice* 41:7–53.

Palfrey, Thomas R. and Howard Rosenthal. 1985. "Voter Participation and Strategic Uncertainty." *American Political Science Review* 79(1):62–78.

Palfrey, Thomas R. and Richard D. McKelvey. 1995. "Quantal Response Equilibria for Normal Form Games." *Games and Economic Behavior* 10(1):6–38.

Przeworski, Adam and John Sprague. 1986. *Paper Stones: A History of Electoral Socialism*. Chicago: University of Chicago Press.

Riker, William H. and Peter C. Ordeshook. 1968. "A Theory of the Calculus of Voting." *American Political Science Review* 62(1):25–42.

Roemer, John. 1997. "Politico-Economic Equilibrium when Parties Represent Constituents: The Unidimensional Case." *Social Choice and Welfare* 14:479–502.

Roemer, John. 2001*a*. *Political Competition*. Cambridge, MA: Harvard University Press.

Roemer, John. 2001*b*. *Political Competition: Theory and Applications*. Cambridge, MA: Harvard University Press.

Schuessler, Alexander. 2000. *The Logic of Expressive Choice*. Princeton, NJ: Princeton University Press.

Seligson, Mitchell A., Annabelle Conroy, Ricardo Córdova Macías, Orlando J. Pérez and Andrew Stein. 1995. Who Votes in Central America? A Comparative Analysis. In *Elections and Democracy in Central America, Revisited*, ed. Mitchell A. Seligson and John A. Booth. Chapel Hill: University of North Carolina Press chapter 7, pp. 151–182.

Simon, Herbert A. 1957. *Models of man, social and rational: mathematical essays on rational human behavior in a social setting*. New York: John Wiley & Sons.

Thelen, Kathleen and Sven Steinmo. 1992. Historical institutionalism in comparative politics. In *Structuring politics: Historical institutionalism in comparative analysis*, ed. Kathleen Thelen Sven Steinmo and Frank Longstreth. Cambridge: Cambridge University Press chapter 1, pp. 1–32.

Tullock, Gordon. 1971. "The Paradox of Revolution." *Public Choice* 11:89–99.

Turocy, Theodore. 2005. "A Dynamic Homotopy Interpretation of the Logistic Quantal Response Equilibrium Correspondence." *Games and Economic Behavior* 51(2):pp. 243–263.

Wittman, Donald. 1983. "Candidate Motivation: A Synthesis of Alternative Theories." *American Political Science Review* 77(1):142–157.

The manufacturer's authorised representative in the EU is Springer Nature Customer Service Centre GmbH, Europaplatz 3, 69115 Heidelberg, Germany. If you have any concerns regarding our products, please contact ProductSafety@springernature.com

Printed and bound by CPI Group (UK) Ltd, Croydon, CR0 4YY
22/09/2026
02221064-0015